FAX:

Digital Facsimile Technology & Applications

The Artech House Telecommunication Library

Vinton G. Cerf, *Series Editor*

Digital Cellular Radio by George Calhoun

E-Mail by Stephen A. Caswell

Introduction to Telecommunication Electronics by A.Michael Noll

Introduction to Telephones and Telephone Systems by A. Michael Noll

Integrated Services Digital Networks by Anthony M. Rutkowski

Digital Signal Processing by Murat Kunt

Principles of Secure Communication Systems by Don J. Torrieri

Digital Image Signal Processing by Friedrich Wahl

Advances in Computer Systems Security: 3 volume set, Rein Turn, ed.

Techniques in Data Communications, by Ralph Glasgal

Measurement of Optical Fibers and Devices by G. Cancellieri and U. Ravaioli

Codes for Error Control and Synchronization by Djimitri Wiggert

Telecommunication Systems by Pierre-Girard Fontelliet

Optical Fiber Transmission Systems by Siegried Geckeler

Traffic Flow in Switching Systems by G. Hebuterne

Optimization of Digital Transmission Systems by K. Trondle and G. Soder

Quality Measures and the Design of Telecommunications Systems by John H. Fennick

Troposcatter Radio Links by G. Roda

Innovation in Internetworking, Craig Partridge, ed.

Principles of Signals and Systems: Deterministic Signals by B. Picinbono

Television Technology: Fundamentals and Future Prospects by A. Michael Noll

Voice Teletraffic System Engineering by James R. Boucher

Advances in Fiber Optics Communications, Henry F. Taylor, ed.

Terrestrial Digital Microwave Communications, Ferdo Ivanek, ed.

Jitter in Digital Transmission Systems by Patrick R. Trischitta and Eve L. Varma

FAX:

Digital Facsimile Technology & Applications

Kenneth R. McConnell
Dennis Bodson
Richard Schaphorst

Artech House

Library of Congress Cataloging-in-Publication Data

McConnell, Kenneth R.
Fax : digital facsimile technology and applications / Kenneth R. McConnell, Dennis Bodson, Richard Schaphorst.
p. cm.
Bibliography: p.
Includes index.
ISBN 0-89006-310-9 : $60.00
1. Facsimile transmission. I. Bodson, Dennis. II. Schaphorst, Richard. III. Title.
TK6710.M33 1989

621.382'35--dc20
89-6602
CIP

ARTECH HOUSE, INC.
685 Canton Street
Norwood, MA 02062

International Standard Book Number: 0-89006-310-9
Library of Congress Catalog Card Number: 89-6602

10 9 8 7 6 5 4 3 2 1

Dedication

To Austin G. Cooley,
one of the prime contributors to the development of the facsimile art,
making possible today's highly successful Group 3 technology.

Contents

Preface		xi
CHAPTER 1 HISTORICAL BACKGROUND		1
1.1	Historical Review of Facsimile	1
	1.1.1 Facsimile Development Problems	4
	1.1.2 The Telephone Era of Facsimile	8
1.2	CCITT Fax Standards	21
	1.2.1 Group 1 Facsimile	21
	1.2.2 Group 2 Facsimile	21
1.3	Background of Digital Facsimile	23
CHAPTER 2 GROUP 3 FACSIMILE		25
2.1	The Group 3 Standard	25
2.2	Group 3 Features	25
2.3	Architecture	28
2.4	Resolution	29
2.5	Compression Coding Overview	31
	2.5.1 Modified Huffman Code	31
	2.5.2 Modified Read Code	32
2.6	Modem	38
2.7	Protocol	40
	2.7.1 Facsimile Call Placement	40
	2.7.2 Signaling For Calling Party Sends	42
	2.7.3 Optional and Nonstandard Facilities Call	46
	2.7.4 Polling Called Party	46
	2.7.5 Handshake Signal Formats	49
	2.7.6 DIS Signal	49
2.8	Group 3 Enhancements	52
	2.8.1 Error Concealment Techniques	52
	2.8.2 Group 3 Error Control Option	57
	2.8.3 Dither Coding	62
	2.8.4 RS-232 Interface	69

2.8.5	High Resolution	71
2.8.6	Small Page Size	75
2.8.7	Nonstandard Operation	77
2.8.8	Group 3 with Group 4–like Features	78
2.8.9	Fax-Phone Switch Boxes	79
2.9	Computer-Based Facsimile	80
2.9.1	PC-Fax Cards	80
2.9.2	PC-Fax Controllers	83
CHAPTER 3	GROUP 4 FACSIMILE	85
3.1	The Group 4 Standard	85
3.1.1	The Seven-Layer OSI Model	86
3.1.2	Resolution	94
3.1.3	Three Classes of Group 4 Facsimile Aparatus	94
3.1.4	Basic and Optional Coding Schemes	96
3.2	Network-Related Requirements	96
3.2.1	Public Switched Telephone Network	97
3.2.2	Circuit Switched Public Data Networks	97
3.2.3	Packet Switched Public Data Networks	97
3.2.4	Integrated Services Digital Network (ISDN)	97
3.3	Mixed Mode	98
CHAPTER 4	SCANNING TECHNOLOGY	101
4.1	CCD Chip Scanners	101
4.2	Contact Image Sensor Scanners	103
4.2.1	Optics CIS	104
4.2.2	Full-Contact CIS	106
CHAPTER 5	PRINTING TECHNOLOGY	107
5.1	Thermal Paper Fax Recording	107
5.1.1	Thermal Print Head Operation	107
5.1.2	Pros and Cons of Thermal Fax	108
5.2	Plain Paper Fax Recording	109
5.2.1	Thermal Transfer	109
5.2.2	Laser Fax Xerographic Recorder	110
5.2.3	LED Linear Array Recorder	111
5.2.4	Liquid Crystal Shutter (LCS) Printer	113
CHAPTER 6	ADVANCED CODING OF BINARY IMAGES	115
6.1	Overview	115
6.2	Input Spatial Filtering	121
6.3	Interpolation	123
6.4	Pattern Recognition	127
6.5	Advanced Group 4 Coding for Binary Images	131
CHAPTER 7	CODING GRAY SCALE AND COLOR PICTURES	135
7.1	Overview	135

7.2	Pulse Code Modulation (PCM)		137
7.3	Predictive Coding		140
7.4	Transform Coding		146
7.5	Vector Quantization		150
7.6	Bit Plane Coding		154
7.7	Color Facsimile		154
CHAPTER 8 FAX TEST CHARTS			163
8.1	Test Charts		163
	8.1.1	IEEE Chart 167A	163
	8.1.2	CCITT Chart No. 1	165
	8.1.3	CCITT Chart No. 2	166
	8.1.4	CCITT Chart No. 3	166
	8.1.5	CCITT Group 3 Test Images	166
CHAPTER 9 THE MARKETPLACE			177
9.1	The Japanese Fax Monopoly		177
9.2	The Development of Fax Markets		179
9.3	Group 3 Equipment in the Marketplace		183
	9.3.1	Hints to Fax Buyers	183
	9.3.2	Hints to Fax Users	185
	9.3.3	Applications for Fax	187
	9.3.4	Low-End Fax	189
	9.3.5	Desk Top Integrated Fax-Phone	191
	9.3.6	Fully Featured (Standard) Fax	191
	9.3.7	Enhanced Features	194
	9.3.8	Cellular Mobile Fax	196
	9.3.9	Small Copy A5 and A6 Fax	197
	9.3.10	PC-Fax Cards	197
	9.3.11	Junk Fax Mail —	198
9.4	Group 4 Equipment in the Marketplace		199
9.5	Facsimile Services		200
	9.5.1	Bureaufax	200
	9.5.2	PTT FAXPOST	201
	9.5.3	Hotelecopy	204
	9.5.4	Additional Fax Services	204
	9.5.5	DHL Worldwide Express	205
	9.5.6	MCI Fax	205
	9.5.7	Electronic Mail with Group 3 Fax Delivery	206
	9.5.8	The Future	206
Glossary of Facsimile Terms			209
Bibliography			217
Index			219

Preface

How did fax, long the stepchild of electronic communication, quite suddenly become the most popular of today's information systems? Not before a century of experimentation in the quest for a practical, efficient method of sending written and printed information did this occur. Names like Morse, Marconi, and Bell are universally recognized, but now that fax is "in," will the name of Alexander Bain come out of obscurity?

Fax is short for facsimile, which is known as *telefax* in some countries. It is the method by which a page (printed, written, or photographic) is converted into electrical signals, sent quickly over a telephone line, and recorded as a copy. Fax works like a remote office copier. The image received can alternatively be displayed on a computer screen (called *soft copy*).

Until the invention of the telegraph and facsimile, the ability to communicate beyond voice range was limited to smoke signals, beats of a drum, and sending a written message by courier. Then, with the discovery of how to code and send electric currents over wires, the courier was replaced. This started a communication revolution, extending the range of the human voice and the human eye around the world and beyond. The rapid sending of text and images by fax, much faster than mail or courier services, is essential to business today.

Actually, facsimile was invented in 1843, but was little used for the next one hundred years. The fax growth rate of about 30% per year started in 1970, but was hardly noticed from the small base of less than 20,000 units in North America. Fax did not become a serious contender for business communication until the 1980 CCITT international standards for digital facsimile. Thus began the explosion in the numbers of new fax units. The additional units installed doubled in 1987, and again in 1988, to make a worldwide population of eight million Group 3 digital fax units by the end of 1988.

Fax is now the method of choice for business text and image communication. Even small businesses must have a fax machine. Sending one- or two-page documents instantly by fax can be cheaper than the 25 cents for later delivery by the

US Postal Service. Fax is more user-friendly than office copiers or computers, a prime reason for its success. Fax is telephone mail for business documents with "instantaneous" low-cost delivery anywhere — across town, across the country, or overseas.

Acknowledgements

The authors wish to thank the following for their helpful information and advice: Esther Bernstein, Austin G. Cooley, Lester M. Davis, Michel Didier, Sydney Fisher, Norman K. Hopkins, Steve Perschau, Allan Pugh, Neil Randall, Donald J. Ryan, John R. Shonnard, George M. Stamps, Cees F. M. Tellings, Stephen Urban, Bill Webb, Pierre-André Wenger, and Dr. M. Zennaki.

We are also indebted to Rita Bodson, Priscilla McConnell, and Patricia Schaphorst for the tireless assistance in research of material for the book, suggestions for writing style and content, and in the endless editing required.

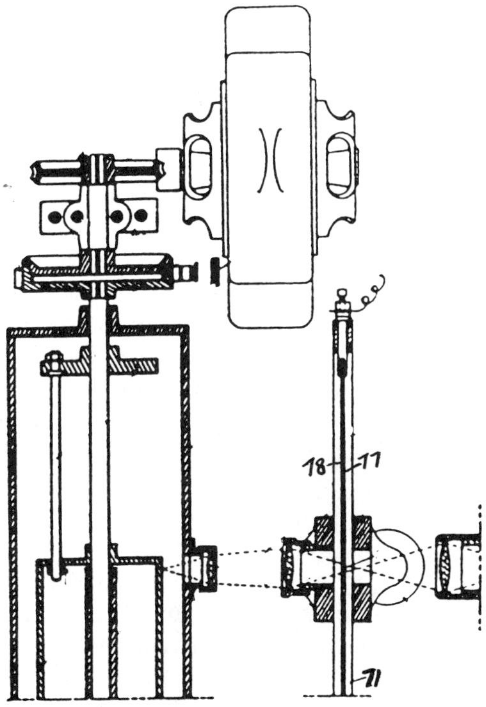

Figure 1.1.7 Korn's telephoto receiver.

of the facsimile art, leading toward today's digital fax. Other significant fax systems were being developed at the same time.

In 1924, higher speed fax systems developed by Western Union, RCA, and AT&T were the next step in transmission of pictures in the United States. At higher speeds, envelope-delay distortion and other distortions in the telephone lines degrade the received picture quality. Envelope-delay distortion is produced by the slightly longer transit time over the telephone line for frequencies near the edge of the voice band. In many cases, it causes the quality of the received facsimile picture to be unacceptable. As the demand for telephone service increased, the older open-wire telephone lines were replaced by cables, which handled more voice circuits.

Cable telephone lines have high shunt capacitance between the wires, causing the signal loss to increase rapidly with frequency. This severely limits the distance between repeater amplifiers. Loading coils added in series with each wire converted the line into a low-pass filter, reducing transmission loss and improving the quality of voice transmission. These loading coils, however, introduced most of the envelope-delay distortion. This problem and many other telephone line impairments were solved by AT&T when telephotography, its newspaper picture facsimile transmission service, was developed.

Dr. Herbert Ives is credited for the telephone line equalization system used in telephotography. It was determined that the envelope-delay distortion between sending and receiving units should be no more than ±300 microseconds (μs). Special filters were designed to add propagation delay in the center of the voice band, making the delay more nearly constant over the range required for sending facsimile signals (1000 to 2600 Hz). These filters were matched to each leg of the telephone circuit to compensate for the envelope delay of that particular section. Amplitude-versus-frequency filters were selected in a similar manner. Although the "private lines" were made from the same lines as used for regular telephone service, only a small percentage of them were conditioned for telephotography. Lines were selected to avoid disturbances that would affect the picture quality. These disturbances included abrupt variations in line loss, envelope-delay distortion, noise, echoes, and crosstalk. Before sending, a key switch was operated on a simplex dc control circuit to set the network of other telephoto stations to receiving. The switch also locked out the echo suppressors and the stepping type of attenuation regulators.

Facsimile was the only high-speed data transmission service in existence at that time. The overall telephone line performance specifications set for facsimile were almost identical to those for data transmission over a voice channel adopted about 20 years later.

The AT&T telephoto equipment formed the basis of today's private line newsphoto systems (Associated Press, United Press International, and Reuters). The AT&T equipment was built like a lathe with a rotating drum and leadscrew. It was later improved with the picture size increased from 4¼ × 6½ inch (108 × 165 mm) to 11 × 17 inch (279 × 432 mm) and sold to Associated Press to initiate their Wirephoto™ service for newspapers. Resolution was 100 lines per inch with a drum speed of 100 rpm for sending at one inch per minute.

Vestigial sideband amplitude modulation was used with a carrier frequency of 2400 Hz. A dc shunt-wound motor drove the drum directly with a generator at 300 Hz, operating in a servo loop. The 300 Hz tuning fork was made of nickel-chromium steel for a small temperature-frequency coefficient. It was operated in an oven held to 50 ± 0.1 degree Celsius. The frequency was maintained to an accuracy of a few parts per million by manual adjustment as necessary. A transmitting and receiving system took a room full of equipment.

Newsphoto services such as Associated Press, Reuters, and United Press International have employed large private line networks to distribute pictures worldwide. The telephone channels used have similar specifications to those devised for this system. Out-of-town pictures printed in newspapers usually have been sent by facsimile from almost anyplace in the world. Thousands of newspapers receive these pictures at the same time. With the newsphoto facsimile units, each picture element (pixel) has full gray scale range, whereas office facsimile units have only a black or white picture element (pel). Simulated gray scale is provided by clustering dots for a somewhat crude gray scale representation at low resolution.

Associated Press now uses a direct-broadcast satellite (DBS) channel operating at 9,600 bits per second (b/s). Resolution has been increased to 170 lines/inch for an 8 × 10-inch picture, sent in nine minutes. Eight bits are sent per pixel, for 256 gray scale shades. A helium-neon laser spot is swept across the recording paper by a mirror galvanometer, exposing dry silver photographic paper. The image is developed and fixed by a thermal unit inside the fax machine.

United Press International has an electrostatic recorder. Each recorded pixel is first written with a full width charge corresponding to black over a rhombic-shaped area that is the height of the recording line. A single stylus mounted on a belt moves across the page to apply the recording charge pattern. If this pixel should be black, nothing more will be done. For half gray, the charge is neutralized over half the spot width, or, for white, over the entire spot. The lightest gray will be a bar that is 1/64th the width of the pixel. At normal viewing distances, it is difficult to tell the difference between the fax recording and a photographic print.

An electronic picture desk was installed at some newspaper offices. This unit stores the pictures in memory as they are received from the newsphoto network. An editor can then view them on a CRT screen and select those which he wants. These pictures are edited electronically in the picture desk and printed for use in the newspaper.

One of the Western Union facsimile developments resulted in a small-copy office fax unit called Deskfax. About 50,000 units were made to send telegrams between the customer's office and the Western Union office. Their use was in customer offices where the volume of telegram business was not sufficient to have a send-receive teleprinter with a trained operator. This successful application disappeared with the demise of the telegram.

In the fall of 1934, Associated Press (AP) contracted for a wirephoto system from AT&T. The *New York Daily News* contributed $150,000 and several other large newspapers put up proportionate amounts. The *New York Times* was not told about the new venture even though the publisher, Adolph S. Ochs, was an AP director. Ochs, quite ill by then, was greatly concerned that the new service would put the Times Wide World Photos out of the photographic business. His son-in-law, Arthur Hays Sulzberger, wanted to outdo AP with a portable photo-facsimile device that could be operated over regular voice telephone lines rather than the expensive AP system. Although Ochs did not think it could be done, he gave Sulzberger permission to try.

Sulzberger contacted Austin G. Cooley who had worked on facsimile for fifteen years and developed a semiportable facsimile system. Cooley was given the job of trying to develop a fully portable unit for test. Fifty-nine work-intensive days later, he had one.

Prior to completion of the new equipment, Sulzberger talked with the head of AT&T, Walter Gifford, and asked, "Does it make any difference to the telephone company if a telephone user talks English or French?"

Gifford said all conversation could be used if it was not profane.

"Then a telephone user can talk French or even Turkish, and that would be all right?"

Gifford nodded.

"How about Turkish or gibberish — that still all right?"

Gifford said, "Arthur, what are you getting at?"

"This: since it doesn't matter what language goes over a telephone, it should make no difference to you if we take a picture and transform its light values into sound values, send them over ordinary telephone lines, and then retranslate them into light values at the other end of the line."

He thought a while before his reply. Then he said, "No, I don't think that would concern us at all."

An earlier report had assured him that no picture transmission would work for more than 100 miles except on the expensive, specially selected and conditioned wirephoto circuits set up by AT&T.

Thus it became permissible to send facsimile signals over the regular switched telephone network, provided that the facsimile equipment was not electrically connected to the telephone company's equipment. Use was limited to newspapers and public services such as police.

The tests with Cooley's wirephoto equipment were successful. In 1935, Wide World Wire Photos (later called Times Facsimile Corporation) was formed, with Cooley running it, to build and operate facsimile equipment for the *New York Times*.

Experiments with Cooley's system were quietly going on in San Francisco at the time that the US Navy's dirigible *Macon* exploded off the California coast on February 12, 1935. Wide World photographers, present when the survivors were brought in to the harbor, took pictures and faxed them to New York. The first publication of any photographs transmitted by this method was carried by the *New York Times* in the account of the *Macon* disaster. The "W.W.W. Photo" credit line shocked telephone company executives and AP wirephoto customers. They found it difficult to believe that good quality pictures could be sent over regular, unconditioned, long-distance telephone lines instead of the $3 million special wirephoto circuits. It then cost no more to send a picture than to talk over the same telephone line.

Instead of being limited to a fixed picture network with sending access only at certain newspaper locations, pictures hence were sent directly from the location of a news event to the *Times* headquarters in New York. The photographer made a print of his picture for the facsimile operator. He then mounted it on a cylinder of a portable facsimile transmitter about the size of a small suitcase, a radical reduction in size from the room full of equipment required by Associated Press. A regular telephone call was placed to the *Times* picture darkroom from any telephone, including a coin-operated one where no ac power was available. A six-volt automobile battery served as the power source for the facsimile transmitter. An inductive coupling unit, slipped over the telephone bell box, connected the facsimile unit to the telephone line without a wired connection to it.

Long-distance telephone operators were a problem, however. They had been instructed not to leave a circuit connected when the customer was finished or something went wrong with the call. As a result, they often monitored the call to see when it was done. When they heard the facsimile signal gibberish, the call was quickly disconnected. On pay phone calls, the fax user had to convince the long-distance operator not to monitor and that the user would pay even if the call took half an hour.

Opening the switched voice telephone network to fax transmissions, when overseas lines became available, provided the channel necessary to access fax units anywhere in the world by merely making a telephone call. Without this access, Group 3 fax would still be a small, insignificant service.

At the start, it was difficult to fabricate facsimile equipment because few of the parts needed were available. This was still the case in 1935 when Times Facsimile Corporation (TFC) began operations. Many of the mechanical, electromechanical, and electrical parts required were unique and were pushing the technological limits for precision. Because there was no established source, it was necessary for the facsimile manufacturer to design and make them, or to arrange with companies that had the background and expertise to develop what was needed.

There was no motor that would fill the performance specifications which TFC needed for its portable facsimile units. To achieve the rotational stiffness needed for high-quality recording, an 1800 Hz, 1800 rpm, 1/100 horsepower synchronous motor was designed and manufactured by TFC. The 50% efficiency was amazingly high. The gap between stator and rotor was only 0.001 inch. Dies for the motor laminations were made in the TFC experimental machine shop. The stamping, heat treating, machining, assembly, and testing were also done internally. Obtaining precision gears, bearings, and machining leadscrews with enough accuracy to prevent picture streaking were serious problems. Even the optics for some of the facsimile systems required unique designs. Audio and power transformers, audio and power filters, solenoids, and other electromechanical components were designed and made at TFC.

There were no frequency standards with the long-term stability required for portable facsimile units. Portable equipment ruled out use of a highly stable oven for constant temperature, and the military required operation over a wide temperature range. John Shonnard, an innovative TFC engineer, patented the production design of a bimetallic, hermetically sealed tuning fork. The resulting oscillator, which had a long-term stability far exceeding available crystal-controlled oscillators, met all military requirements. Hermetic sealing was required to prevent frequency variations with pressure and humidity changes. A stainless steel cylinder holding the tuning fork alone was evacuated to prevent internal pressure changes with temperature. Magnetically coupled drive and pickup coils were mounted outside the cylinder. Earlier attempts using coils inside the cylinder led to long-term frequency instability. Production tuning forks were individually processed and tested over a period of a month to ensure that the tight specifications were met.

When the picture service provided by the Associated Press expanded and improved, Sulzberger, by then publisher of the *Times,* sold Wide World Wire Photos to AP, but kept the facsimile development and manufacturing facilities to build facsimile equipment for the US government during World War II. He planned to dissolve TFC at the end of the war because the *Times* no longer needed its own telephoto equipment.

The US military had become interested in this equipment and wanted a small-sized portable transceiver. The large wirephoto receiver functions were squeezed into the small transmitter case. The unit could now send, receive positives on photographic paper, receive negatives on film, or receive on a direct recording paper. After testing by the Signal Corps, the FX-1B, a 22 × 12 × 10-inch militarized transceiver was designed and produced. (See Figure 1.1.8.)

The FX-1B was one of the first pieces of equipment that successfully passed the rugged Signal Corps Laboratories environmental testing for military equipment. FX-1B was used during the invasion of Normandy to send photographs across the English Channel for directing artillery fire. In the China-Burma war theater, one of the FX-1Bs was jettisoned from an airplane. The case was severely damaged, but after replacement of a small casting by a TFC technician, the unit worked perfectly.

Another application of the FX-1B was sending up-to-date weather maps from Washington, DC, for the service that ferried warplanes to Europe via Presque Isle, Maine, Gander, Newfoundland, Goose Bay, Labrador, and Reykjavik, Iceland. Only a very small portion of a weather chart could be sent at a time because of the 7 × 8⅝-inch picture size.

International standards for weather map fax transmission were set one day by two men because of this problem. After struggling for months with the limitation, US Army Air Force Sergeant Chuck Halbrooks added "modified" to an order for FX-1Bs and slipped it in with other papers awaiting signature. He then visited TFC and told Cooley he would soon receive this order. Halbrooks and Cooley quickly altered the design. The drum diameter was increased from 2.75 inches to 6 inches, using 6-inch aluminum tubing, which was available without special order.

Figure 1.1.8 World War II facsimile transceiver.

This gave 18.85-inch circumference for a usable copy width of 18 inches. The drum length was increased to 12 inches. The sending rate was speeded up by 50%. The 96-lines/inch leadscrew was retained. The new design worked very well and was standardized by the military, with orders for over 1000 units eventually placed.

Halbrooks paid for his action by being transferred to duty in the Far East, but the Army Air Force had a much needed weather fax unit.

The US Army Air Force's use of facsimile for weather map transmission during World War II had set the worldwide standards for weather facsimile. The International Telegraph and Telephone Consultative Committee (CCITT) adopted somewhat different standards, but abandoned them for those set at TFC. After the war, networks were set up for broadcast of facsimile weather charts over telephone lines and by HF radio.

Although production stopped on most military equipment at the end of World War II, the Army, Navy, Air Force, and the Weather Bureau needed many more weather fax units. Associated Press needed lightweight portable wirephoto fax units similar to the FX-1B. In the next few years, new facsimile equipment such as Weatherfax, Stenafax, Policefax, Pressfax, Messagefax, and many other types of specialized units were designed and manufactured. Faced with an expanding need for the TFC capabilities, Sulzberger abandoned his plan for closing down the facsimile operation, and kept it for another fourteen years until Litton Industries bought TFC and took over its operation.

A facsimile edition of the *New York Times* was a journalistic milestone in 1956. A ten-page edition was sent from the *Times* offices in New York to San Francisco and distributed to the delegates at the Republican National Convention. Figure 1.1.9 shows the author, Ken McConnell, in the transmitting room during the operation.

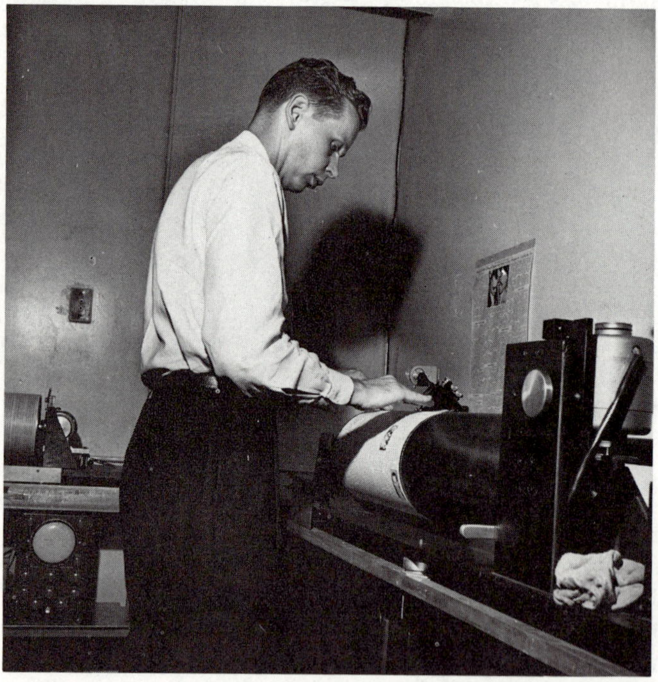

Figure 1.1.9 First Pressfax newspaper.

A special wideband 480 kHz channel provided by AT&T carried the facsimile signals for the one-week duration of the experiment. The *Times* was printed there from the full-page facsimile negatives. The quality of the paper was such that few delegates realized they were looking at a facsimile reproduction. The 200-lines/inch facsimile equipment, specially designed for the experiment, was the first fax that would send a full newspaper page at a time. In an earlier experiment in April 1945, the *Times* had published a four-page edition for the United Nations conference at San Francisco using AP Wirephoto units to send a half page at a time with much lower resolution of 100 lines/inch.

Based on this Pressfax equipment, a much higher resolution system was made for the *Wall Street Journal.* They built a new printing plant without typesetting capabilities in Riverside, California, near Los Angeles. A television channel from San Francisco connected the transmitter there to the receiver in Riverside. Functioning as replacements for typesetting, the Pressfax receivers produced full-page negatives so sharp (1000 lines/inch) that even their own technicians found it impossible to distinguish the papers printed in Riverside from those printed in San Francisco.

Today, newspapers and magazines are printed in plants that may be thousands of miles away from the place where the pages are composed. High-speed, high-resolution facsimile units can send pages up to 22 × 28 inch (two full-sized newspaper pages) over wideband satellite or land lines. Both analog and digital types of facsimile units are in use. Some units require a T-1, 1.34 Mbit or a T-2, 6.34 Mbit channel. Resolutions of 800 to 1000 lines/inch are commonly used, but 1800 lines/inch and higher are available. The facsimile recording is made on a film negative or directly on a printing plate. The *Wall Street Journal, New York Times, USA Today,* and other newspapers send to many different printing plants via DBS. A satellite receiving station is located at each printing plant. *USA Today* covers the United States with 32 printing plants, and has additional ones overseas. Color pages are sent by making three or four transmissions of the original color page using color separation filters.

Office facsimile equipment was designed and manufactured in the United States, United Kingdom, France, Germany, and Japan. At first, these were incompatible analog facsimile units that operated on different standards. Each manufacturer touted the advantages of its design and did not want compatibility because that would invite competition for its customers. It was not possible to use the regular telephone lines (now called the *public switched telephone network* or PSTN), so private line networks were utilized. Most businesses either did not know facsimile existed or saw no use for it.

When the Bell System offered a facsimile coupling device with a special analog modem for generating frequency shift signals to send over the PSTN, connection to telephone lines improved. Although the Bell System modem worked quite well, it was not liked. The analog modem, already part of the fax unit, could not be used, and additional components were required to interface with the Bell System modem.

The Electronics Industry Association (EIA) established a technical committee TR-29 on facsimile equipment and systems in the early 1960s. At its meetings the engineers were restricted by their companies from agreeing to a fixed set of specifications with normal tolerances because it would mean their designs would have to be changed. It was not until October 1966 before EIA Standard RS-328, Message Facsimile Equipment for Operation on Switched Voice Facilities Using Data Communication Equipment, was published. This was the first US "standard" on office fax. The following is quoted from RS-328.

> Each manufacturer has designed his facsimile equipment used for message communications to fit the requirements of his customers and the natural characteristics of the types of mechanisms employed when operating on a private line basis. These same equipments were initially used when operation of facsimile equipments on a regular telephone call basis became available. Operation between some of these equipments was not possible due to differences in the facsimile equipments. This Standard specifies machine characteristics which will ensure interoperation.

With this standard, the received copy might be stretched from the original size or could have parts near the edge of the page missing, but, for the first time, diverse fax units were able to communicate. Each manufacturer stayed within the wide tolerances required to get the standard, but did not convert to the "recommended for eventual standard" items. Clearly, there was an acute need for a facsimile standard to one set of performance specifications instead of interoperability.

The fax manufacturers used acoustic coupling to and from the telephone handset, avoiding the Bell System analog modem. Acoustic coupling had some problems with room noise and distortion caused by the carbon granules of the telephone transmitter.

Compatibility improved as companies like Magnavox used the EIA-recommended fax equipment specifications on new equipment designs. Because the Magnavox units were marketed through Xerox Corporation, these standards were kept when that firm started manufacture of its own units. Other companies such as Graphic Sciences and Minnesota Mining and Manufacturing kept their own different standards, even after the CCITT Group 1 facsimile standard was adopted.

The 1967 Carterphone decision allowing direct connection to the PSTN was a substantial improvement for fax. At last the fax signals could be electrically connected to the regular telephone lines (PSTN). The signals, however, still had to pass though a telephone company coupling device to isolate and protect the telephone network. About 10 years later fax units were finally permitted to connect directly to the telephone lines. Table 1.1.1 shows technical progress in facsimile and some other communication systems since conception of the first fax unit.

There are too many other developments and contributors to the facsimile art to mention all of them in addition to those shown.

Table 1.1.1. Facsimile Technical Progress 1843—1988

1843	First Facsimile, Bain — Pendulum Type
1844	Telegraph, Morse
1850	Rotating Drum, Bakewell — England
1865	First Commercial Fax, Caselli
1876	Telephone, Bell
1902	Optical Scan, Dr. Korn — Germany
1906	News Photo Service, Dr. Korn — Munich to Berlin
1907	Vacuum Tube (Audion), De Forrest
1915	Telephone Repeaters — Long Distance
1917	Teletype — AT&T
1924	Photofax in US — AT&T, RCA, and WU Bell 3002 Envelope Delay Specs — VSB
1926	RCA Transatlantic Newsphoto Service by Radio
1928	Photo Recording without Optics, Cooley
1934	Associated Press Photofax News Picture Service
1936	Drum and Helix Recorder — Electrolytic Paper, Dr. Hogan
1936	Drum and Helix Recorder — Carbon Paper, RCA — Artzt
1936	Dry Direct Recording Paper, Western Union — Teledeltos
1936	*New York Times* Manufactures Fax, Times Facsimile — Cooley
1943	Signal Corps Fax FX-1B, Times Facsimile
1943	Timefax — Direct Recording Dry Paper
1945	TV — Regular Broadcast in USA
1948	Western Union Deskfax — 40,000 Units
1956	Acoustic Coupling on PSTN Lines in USA
1958	Stewart-Warner Datafax
1959	Telautograph Quickfax
1960	EIA Committee TR-29 Starts Standard for Fax
1965	EIA Issues Fax Standard RS-328
1965	Magnavox Telecopier — Xerox Markets It First US Large Production Office Fax

Table 1.1.1. Facsimile Technical Progress 1843—1988 (Continued.)

1967	"Carter Phone" Decision — Direct Connection to PSTN
1967	Xerox Manufactures Fax
1967	DEX 1 — Graphic Sciences Fax
1968	CCITT Group 1 Fax Recommendation
1968	Scanatron Message Fax
1969	Digital Fax Starts, DACOM — Weber Run Length Code
1970	Xerox 400
1972	3M Enters Fax Market — VRC 600 Manufactured by MGCS (Matsushita)
1973	3M 603 — VSI/MGCS KD-211
1974	RapiFax 100 Digital Fax — DACOM
1974	QWIP 1000 — EXXON
1974	Xerox 410
1975	Xerox 200
1975	100,000 Fax Units in USA
1976	QWIP 1200
1976	CCITT Group 2 Recommendation
1976	CCITT Starts Group 3 Fax Standards
1977	3M 9600 — "Near Group 3" Digital Fax
1978	Panafax UF-20 and UF-320 Digital Fax
1978	DEX 5100 "Near Group 3" Digital Fax
1979	Southern Pacific — DMS 2000 "Near Group 3" Digital Fax Manufactured by Hitachi
1979	Telautograph Omnifax "Near Group 3" Digital Fax Manufactured by OKI — Thermal Recording
1979	NEC 6200 "Near Group 3" Digital Fax
1980	CCITT Group 3 Recommendation
1984	CCITT Group 4 Recommendation
1988	Error-Free Group 3
1988	Small Format (A5 and A6) Group 3

1.2 CCITT FAX STANDARDS

The CCITT is part of the International Telecommunication Union (ITU), a specialized agency of the United Nations. The CCITT studies needs for new types of telecommunication equipment and services, and then develops recommendations which have the effect of standards for international communication. The CCITT works in four-year cycles, and it publishes the new and revised recommendations once every four years. The current period ended in December 1988 and the results were published in the so-called Blue Book. These recommendations for office facsimile units are listed in Table 1.2.1.

1.2.1 Group 1 Facsimile

Standards for Group 1 made possible the consideration of more generalized business use of fax. Group 1 provided compatibility between fax units outside North America, but those in North America were different and could not communicate with Group 1 fax units. Most of the specifications were the same, except for the black and white frequencies, as shown in Table 1.2.1.1.

The US engineers thought that 2400 Hz black and 1500 Hz white worked better over their phone lines than the 2100 Hz black and 1300 Hz white of Group 1. The US salespeople mistakenly called the US fax Group 1. Although the Group 1 units generally worked better than the earlier office fax units, they were still unreliable six-minutes-per-page analog units, which gave fuzzy copy and required much manual attention. Many units offered an optional mode that moved the original page through in four minutes instead of six minutes. This produced lower resolution in the paper feed direction for even lower copy quality. This option was thus a poor attempt to increase the transmission speed.

1.2.2 Group 2 Facsimile

Users who had accepted the quality of Group 1 needed a faster system. Both Xerox and Graphic Sciences developed and marketed new fax units with three minutes per page instead of six minutes, or two minutes per page instead of four minutes. Xerox had a frequency modulation (FM) scheme and Graphic Sciences had an amplitude modulation (AM) scheme. Again, there were no standards.

Then, engineers from competing companies on the EIA technical committee on facsimile, in cooperation with engineers from the United Kingdom and other countries, made tests to determine the best system. As a result of this work, the

Table 1.2.1. CCITT Fax Recommendations

Group	*Recommendation*	
1	T.2	Facsimile Apparatus for Document Transmission over the PSTN
2	T.3	Facsimile Apparatus for Document Transmission over the PSTN
3	T.4	Facsimile Apparatus for Document Transmission over the PSTN
2, 3	T.30	Procedures for Document Transmission over the PSTN
	[T.5 is replaced by the following T.500 series recommendations.]	
4	T.563	Terminal Characteristics for Group 4 Facsimile Apparatus
4	T.503	Document Application Profile for Interchange of Group 4 Documents
4	T.521	Communication Profile for BT0 Bulk Document Transfer under T.62 Session Service Environment
4	T.6	Facsimile Coding Schemes and Coding Control Functions for Group 4 Facsimile Apparatus
4	T.62	Control Procedures for Teletex and Group 4 Facsimile Services
4	T.70	Network Independent Basic Transport Services for the Telematic Services
4	T.72	Terminal Capabilities for Mixed Mode Operation
	[T.73 is replaced by the following T.400 series recommendations for: Document Transfer, Access, and Manipulation (DTAM); Office Document Architecture (ODA); and Interchange Format.]	
4	T.411	Introduction and General Principles
4	T.412	Document Architecture
4	T.414	Document Profile
4	T.415	Document Interchange Format (ODIF)
4	T.416	Character Content Architecture
4	T.417	Raster Graphics Content Architecture

Table 1.2.1.1. CCITT Group 1 and Group 2 Fax

	Group 1	*AM.6M*	*Group 2*
Lines per Minute	180	180	360
Modulation	FM	FM	VSB AM/PM [a]
Carrier Frequency			2100 ± 10 Hz
White Signal	1300 Hz	1500	Maximum Carrier
Black Signal	2100 Hz	2400	26 dB Minimum Lower

[a] VSB is vestigial sideband; PM is phase modulation.

CCITT standard for Group 2 fax was developed. The three-minute fax units already in service were not compatible with the new CCITT standard. When all companies switched to the new Group 2 standard, fax units became compatible worldwide for the first time.

The Japanese established efficient facsimile manufacturing facilities, which, combined with low labor costs, allowed them to become preeminent in the design and manufacture of most facsimile equipment. CCITT standards opened worldwide markets for the Japanese equipment. The much lower price of Group 2 fax units allowed them to compete favorably against the expensive private-standard digital facsimile units, which became available in 1974. Even after the Group 3 standards existed, the Group 2 price advantage remained for a few years.

1.3 BACKGROUND OF DIGITAL FACSIMILE

Good copy quality for office facsimile started with digital facsimile, six years before the CCITT digital fax standards existed. The earliest digital fax units used the adaptive run-length coding algorithm patented by Donald Weber. This coding removes redundancy from the page being sent and thus shortens the transmission time. A digital code word represents the number of successive white picture elements along the scanning line before the next black pel. The next code word represents the number of black pels following the white run. The code words are usually much shorter than the number of pels, and so it takes fewer bits to send the page. Single-line coding under this patent was used in Dacom's Rapifax 100.

A three-line technique under the same patent is still used for coding newspaper pages for fax transmission.

The digital modem for the Rapifax 100 was the first modern high-speed modem employing large-scale integration (LSI) to be used successfully on regular telephone lines (PSTN). The modem further reduced transmission time by coding 4 bits at a time, requiring only 1200 baud (changes per second of the signal state on the telephone line) to send 4800 b/s. Digital fax units were sold by Dacom in the United States and later by Kalle in Europe.

When other digital facsimile units were designed, each company had its own proprietary algorithm for compression of the data and could send only to others of its own manufacture. The need for a digital facsimile standard instead of competing and incompatible units soon became evident.

The telegram of Postal Telegraph and Western Union, dominant until the 1940s, was replaced by telex, especially for international communication. In 1980 the CCITT adopted a standard for teletex, designed as a high-speed high technology to replace telex. Facsimile was not taken seriously by many because it was an "inefficient" way to send text, requiring ten times as much data to be sent for the same message. Yet, teletex has been a dismal failure, whereas Group 3 fax has been tremendously successful.

Chapter 2
Group 3 Facsimile

2.1 THE GROUP 3 STANDARD

The success of Group 3 digital facsimile resulted from international cooperation in generating an excellent standard written to fill a well identified need. Fortunately, facsimile engineers worldwide, who had developed the Group 2 analog standard, again cooperated in formulating the CCITT Group 3 digital facsimile standard adopted in 1980. The V.29 modem was intended to be used on four-wire leased telephone channels conditioned for high-speed data transmission. Bell System literature said their regular telephone lines (PSTN) could not be used successfully at 9600 b/s. Nevertheless, digital facsimile units successfully used V.29 with half-duplex on the PSTN, and V.29 was incorporated as an option in the Group 3 facsimile standard.

The standards for Group 3 are Recommendations T.4 and T.30 in the CCITT Blue Book, including revisions completed November 1988. The main revisions of T.4 and T.30 are the addition of an optional error-free method of transmission and for two smaller-page versions of Group 3 fax. Earlier revisions in the Red Book (1984) increased the usefulness of the Group 3 standard without compromising compatibility with Group 3 units made to the original T.4 Yellow Book (1980) recommendations. (See Table 2.1.1.)

2.2 GROUP 3 FEATURES

One of the most important features of the Group 3 standard is its flexibility, allowing individual manufacturers to add their own innovative features without giving up basic compatibility with standard Group 3 units.

Other important features are listed below:

Acceptably High Resolution — The standard resolution of 203 lines/inch horizontally and 98 lines/inch vertically provides quite readable copy quality for most text and the optional fine vertical resolution of 196 lines/inch improves readability of

Table 2.1.1. CCITT Group 3 Fax Specifications

	Standard	*— Optional —*	
Pels per Line	1728	2048	2432
Horizontal			
Pels per Inch	203		
Pels per Millimeter	8		
Vertical			
Pels per Inch	97.80	195.6	
Pels per Millimeter	3.85	7.7	
Scan Width			
Inches	8.46	10.0	11.9
Millimeters	215	255	303
Coding	*Modified Huffman*	*Modified Read*	
Data Modem	V.27(ter)	V.29	
Bits per Second	4800/2400	9600/7200	
Signaling Modem	V.21	V.27(ter)	
Bits per Second	300	2400	
Milliseconds per Recording Line	20	0, 5, 10, 40	

smaller text. Most Group 3 units have this option. Both resolutions are a substantial improvement over Group 1 and 2 fax units which produced readable but fuzzy copy.

Pages Per Minute Speed — Group 3 units send an average page of text in 10 to 30 seconds, plus about 15 seconds for the first page initial handshake. This assumes the telephone line trains at 9600 b/s, as in most cases. The second and following pages of the same fax session do not have this handshake. The time per page depends on the number of black markings and fineness of detail. The more enhanced Group 3 units with memory storage will send at the higher speeds. Group 1 units took about 6.5 minutes for every page if they had automatic operation. A number of these units were manually operated and required an additional one to seven minutes per page of operator time. Pages often needed to be resent because of operator error. Verification of readable copy received and incidental voice conversation added to the inefficiency. Many Group 2 units were automatic and the time of about 3.5 minutes per page was twice as fast as Group 1.

User-Friendly Operation — Group 3 units are very easy to operate and in many cases are simpler to use than an office copier. Reception of fax documents requires no operator attention. The Group 3 unit is left turned on with low standby power and automatically answers an incoming ring. After automatic reception of the fax document, the unit hangs up, releasing the fax line until the next call is received. For sending, most Group 3 units have automatic feed of multipage documents. After placing a telephone call to the receiving fax number, the "send" button is pressed when the answer tone is heard. Nothing more need be done. Many Group 3 units have automatic dialers, making it unnecessary to dial the receiving fax number or to wait for an answer tone. Transmission of documents anywhere in the world can be done by making a phone call and pressing the send button or by pressing only one button in some autodial fax units. The facsimile units then handshake to determine the highest speed possible on this particular phone call. Transmission can be as fast as 10 seconds per page. The operator need not be concerned about the protocols used or modem compatibility. No conversion of signals is needed to cross international borders. Operation is far simpler than sending messages by computer or telex.

Universal Compatibility — Group 3 fax compatibility is similar to telephone compatibility. Although the units are made by different manufacturers and may have different features, all of them can work properly with each other. Any language used for text and any image will be copied in a manner similar to an office copier. It does not matter whether the receiving fax unit makes a laser copy, uses thermal recording, or some other method.

Use of Regular Telephone Lines (PSTN) — The PSTN is available almost anywhere in the world and a new Group 3 fax unit can be installed with instant availability to eight million other Group 3 units, merely by plugging it into a telephone jack. Group 3 fax on the PSTN has no customs delay or censorship when sending documents to foreign countries, contrasted with mail or courier. If the PSTN is not used, a private line voice or digital channel is needed. These networks have very limited coverage. Connection of fax units to these lines is far more difficult and also more expensive.

Step-Down Modems — These modems automatically test the performance characteristics of the telephone connection and select the highest speed available. Automatic equalization to minimize line distortions makes higher speeds available.

Electronic Handshake — Group 3 units have many different features that must be matched between the transmitter and receiver. The electronic handshake takes care of this automatically and selects the fastest transmission possible between the two units.

2.3 ARCHITECTURE

To understand the architecture of a Group 3 fax unit, the system will be broken up into basic blocks for transmitting and receiving functions. (See Figure 2.3.1.)

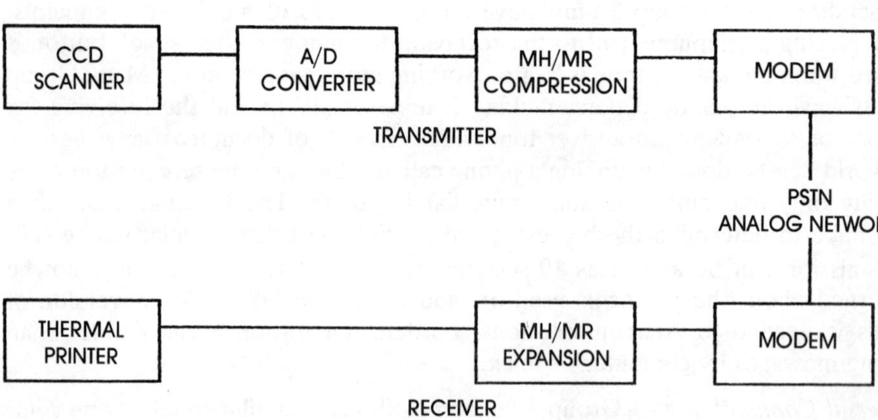

Figure 2.3.1 Group 3 block diagram.

At the transmitter, the charge coupled device (CCD) *scanner* reads the page being sent. Its output signal is a series of pulses. The amplitude of each pulse represents the brightness of the image on each element of the CCD chip. Block A/D converts the signal from analog to digital with one bit per picture element. The MH/MR (*modified Huffman–modified read*) block generates code words containing the picture information in a compressed format that needs only 1/5th to 1/20th as many bits. This compressed digital signal is converted by the *modem* block into an analog signal that can be sent over regular telephone lines (PSTN). A

Group 3 fax unit either sends or receives at one time and some parts serve either sending or receiving functions. There is only one modem block.

The modem at the receiver amplifies the fax signal and converts it back into a digital signal. The MH/MR block now operates in expansion mode to restore a bit stream which represents the black and white pels from the CCD scanner. The *thermal printer* converts the bit stream into a copy of the original page.

To describe in more detail: The scanner reads a very narrow line, 0.01 or 0.005 inch (0.254 or 0.127 mm) high, across the width of the page being sent. The scanning of 1728 pels across the page generates 1728 bits per line. A two-line memory stores adjacent scanning lines. (See Figure 2.3.2.)

Modified Huffman (MH) run-length coding or modified read (MR) two-dimensional coding then compresses the pel information to a small fraction of the number of bits required before coding. This is a source encoding, called redundancy reduction. See Section 2.5 for details of Group 3 coding. A buffer memory stores the output of the MH/MR coder for use by the modem. The modem further codes the digital signal from the buffer memory for transmission as an analog signal over the telephone line. When operating at 9600 b/s, the modem takes 4 bits at a time and represents them as one of 16 different states of *pulse amplitude modulation* (PAM) for transmission as an analog signal. The telephone line has only 2400 baud (changes per second) to convey 9600 b/s. An FCC or other PTT-approved built-in *line connection unit* (LCU) provides a standard miniature jack RJ-11C or other standard connector for direct connection to regular telephone lines (PSTN). Another built-in RJ-11C jack allows use of a telephone for dialing or voice communication.

The same LCU, modem, buffer memory, MH/MR coder, and line memories are used for the receiving function in a half-duplex mode. The modem decodes the received analog fax signal, regenerating the digital signal sent. The MH/MR block then expands these fax data to black-white pel information for printing. The thermal printer has wires spaced at 203/inch touching the temperature-sensitive recording paper. Heat is generated in a small spot on each wire when high current for black marking is passed through it. To mark a black spot, the wire heats from nonmarking temperature (white) to marking temperature (black) and back to white in a few milliseconds.

2.4 RESOLUTION

The standard resolution of 203 lines/inch across the page and 98 lines/inch down the page is adequate for easy readability of text, letterheads, business letters, and memoranda. For printed material from brochures, books, magazines, and newspapers, fine resolution of 196 lines/inch should be used. There are twice as many pels to send, but the time to send is not nearly doubled because compression

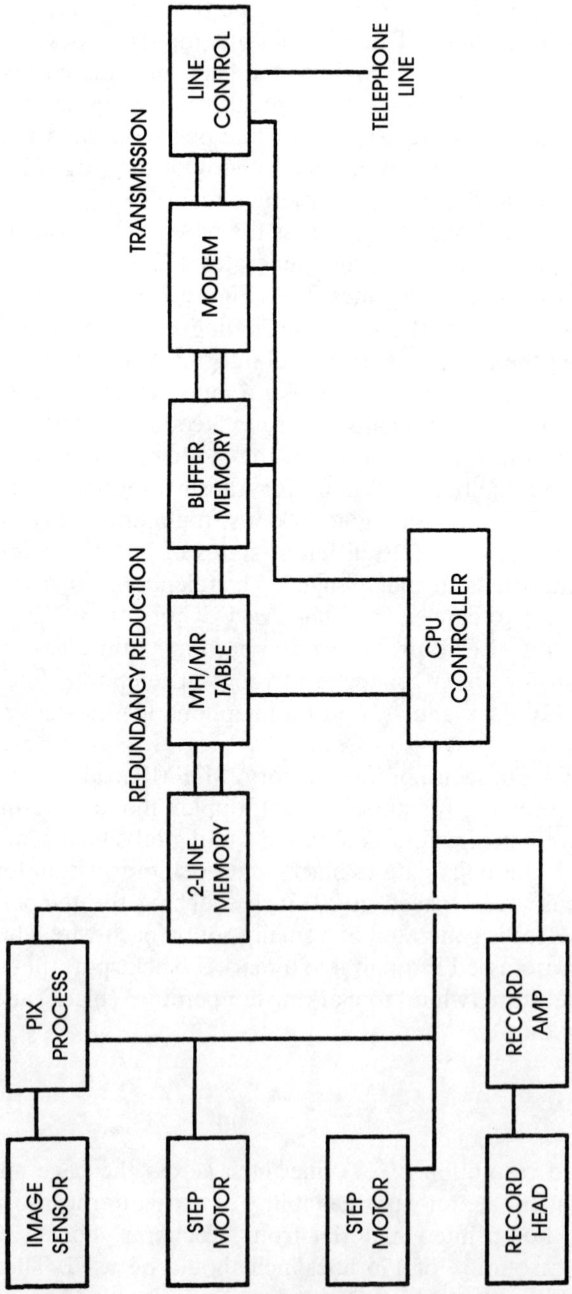

Figure 2.3.2 Group 3 apparatus block diagram.

efficiency is much better with the longer runs. Fortunately, most Group 3 fax units have the higher resolution option.

2.5 COMPRESSION CODING OVERVIEW

Facsimile systems had been handling information in real time as an analog signal. The idea of processing the signal by clocking, quantizing, and storing digital information seemed like science fiction. Scanning and recording systems were mechanical, and it was difficult to mark a straight line down a received page without noticeable jitter (a saw-toothed effect in the recording). A single photosensor was common in the scanner and a single writing point in the recorder. Systems that used two or four scanning or recording spots complicated the design by requiring exact alignment between these spots in addition to the real-time drive problems. Keeping the scanning and recording spots in exact step with the clock was not possible. There was also a drift between the clocks at the transmitter and receiver units. Manual adjustment of the frequency kept the skew within acceptable limits. Digital modems were not available for sending the signals and furnishing a locked-in sampling clock at the receiver. Practical techniques had not been developed for digital storage of the pel information of one or two scanning lines. Quantizing the analog signal of the photographic recording newsphoto systems would require at least 32 gray levels to match the quality of the better photofax units available by 1950. From the information theory later described by Claude Shannon, these analog systems were sending information at about 6600 b/s on voice-type channels.

Technology improved, however. When digital techniques were developed, it became practical to remove redundancy from the page being scanned and restore pel information for recording.

2.5.1 Modified Huffman Code

CCITT Recommendation T.4 specifies that Group 3 fax units must incorporate modified Huffman run-length coding of scanning lines. In a scanning line, a white pel is likely to be followed by a long string of white pels before a black pel is reached. In typed text, the white strings, called *runs,* often continue across the whole page. Each sensor element has one bit of information to represent a black or white pel. Instead of sending a white line across the page as 1728 bits, MH sends a 9-bit code word. This compresses the information by a ratio of 1728/9 = 192. There are many white runs that are much shorter, and these have different code words.

A total of 1728 code words would be needed to cover all possible run lengths across the page. To make a shorter code table, the runs are grouped in multiples of 64 pels in the "make-up code" table, and short runs of 0 to 63 are placed in a separate "terminating code" table. Only 92 binary codes are then needed for any white run-length of 0 to 1728 pels by sending the make-up code followed by the terminating code. Black runs are coded using 92 different codes. (See Figures 2.5.1.1 and 2.5.1.2.)

To determine which code words should be shortest, eight test pages with different languages and content were scanned. (See Figure 8.1.5.1.) Each black and white run length and the number of times it occurred were tabulated. The shortest code words were assigned to the run lengths occurring most often. This is an application of the coding principle first used in 1844 by Morse to send telegraph signals efficiently.

Figure 2.5.1.3 shows a scanning line with black and white runs and the code words for these runs.

The code word for a 2-pel white run is 4 bits, taking twice as long to send as uncompressed. Because a 2-pel white run is infrequent, a short code word is saved for a longer run, which happens more often. A white run of 585 pels takes only 13 bits for a compression of 45. At the end of each coded scanning line an "end of line" (EOL) code of 000000000001 is sent. The long sequence of 0s makes it easy to detect from the run length coded words. At the receiving fax unit, the decoded runs for one line are added to check whether there is an error in the line. If the total number of pels is not 1728, one or more errors have occurred. For wider copy fax units, the total number would be 2048 or 2432.

2.5.2 Modified Read Code

Most information on a page being sent has a high degree of vertical correlation. Modified read is a *re*lative *ad*resssing code which uses this vertical correlation to achieve a higher compression than modified Huffman coding. With two adjacent scanning lines on a page, black pel runs of the second line often start directly under a black pel run in the first line and *vertical mode* coding uses only one bit to indicate this condition. In those areas of the page where there is little vertical correlation, *pass mode* or *horizontal mode* is used.

Line memory in the fax units stores each pel of two or more scanning or printing lines so the vertical coding can be performed. The first scanning line in memory is called the *reference line.* It has already been coded, but the location of each black pel of this line is in memory as a reference for the next scanned line called the *coding line.* (See Figure 2.5.2.1.)

Looking at the reference line and coding line, move from left to right along the pair of lines until a pel on either line changes its color (from white to black or from black to white). This is called a *changing pel.* If a pair of *changing pels* on the adjacent lines match or are within three pels, vertical mode coding is selected. (See Figure 2.5.2.2.)

Terminating codes

White run length	Code word	Black run length	Code word
0	00110101	0	0000110111
1	000111	1	010
2	0111	2	11
3	1000	3	10
4	1011	4	011
5	1100	5	0011
6	1110	6	0010
7	1111	7	00011
8	10011	8	000101
9	10100	9	000100
10	00111	10	0000100
11	01000	11	0000101
12	001000	12	0000111
13	000011	13	00000100
14	110100	14	00000111
15	110101	15	000011000
16	101010	16	0000010111
17	101011	17	0000011000
18	0100111	18	0000001000
19	0001100	19	00001100111
20	0001000	20	00001101000
21	0010111	21	00001101100
22	0000011	22	00000110111
23	0000100	23	00000101000
24	0101000	24	00000010111
25	0101011	25	00000011000
26	0010011	26	000011001010
27	0100100	27	000011001011
28	0011000	28	000011001100
29	00000010	29	000011001101
30	00000011	30	000001101000
31	00011010	31	000001101001
32	00011011	32	000001101010
33	00010010	33	000001101011

Figure 2.5.1.1 Modified Huffman terminating codes.

Make-up codes

White run lengths	Code word	Black run lengths	Code word
34	00010011	34	000011010010
35	00010100	35	000011010011
36	00010101	36	000011010100
37	00010110	37	000011010101
38	00010111	38	000011010110
39	00101000	39	000011010111
40	00101001	40	000001101100
41	00101010	41	000001101101
42	00101011	42	000011011010
43	00101100	43	000011011011
44	00101101	44	000001010100
45	00000100	45	000001010101
46	00000101	46	000001010110
47	00001010	47	000001010111
48	00001011	48	000001100100
49	01010010	49	000001100101
50	01010011	50	000001010010
51	01010100	51	000001010011
52	01010101	52	000000100100
53	00100100	53	000000110111
54	00100101	54	000000111000
55	01011000	55	000000100111
56	01011001	56	000000101000
57	01011010	57	000001011000
58	01011011	58	000001011001
59	01001010	59	000000101011
60	01001011	60	000000101100
61	00110010	61	000001011010
62	00110011	62	000001100110
63	00110100	63	000001100111

Figure 2.5.1.1 (continued).

Make-up codes

White run lengths	Code word	Black run lengths	Code word
64	11011	64	0000001111
128	10010	128	000011001000
192	010111	192	000011001001
256	0110111	256	000001011011
320	00110110	320	000000110011
384	00110111	384	000000110100
448	01100100	448	000000110101
512	01100101	512	0000001101100
576	01101000	576	0000001101101
640	01100111	640	0000001001010
704	011001100	704	0000001001011
768	011001101	768	0000001001100
832	011010010	832	0000001001101
896	011010011	896	0000001110010
960	011010100	960	0000001110011
1024	011010101	1024	0000001110100
1088	011010110	1088	0000001110101
1152	011010111	1152	0000001110110
1216	011011000	1216	0000001110111
1280	011011001	1280	0000001010010
1344	011011010	1344	0000001010011
1408	011011011	1408	0000001010100
1472	010011000	1472	0000001010101
1536	010011001	1536	0000001011010
1600	010011010	1600	0000001011011
1664	011000	1664	0000001100100
1728	010011011	1728	0000001100101
EOL	000000000001	EOL	000000000001

Figure 2.5.1.2 Modified Huffman make-up codes.

Run length (black and white)	Make-up codes
1792	00000001000
1856	00000001100
1920	00000001101
1984	000000010010
2048	000000010011
2112	000000010100
2176	000000010101
2240	000000010110
2304	000000010111
2368	000000011100
2432	000000011101
2496	000000011110
2560	000000011111

Figure 2.5.1.2 (continued).

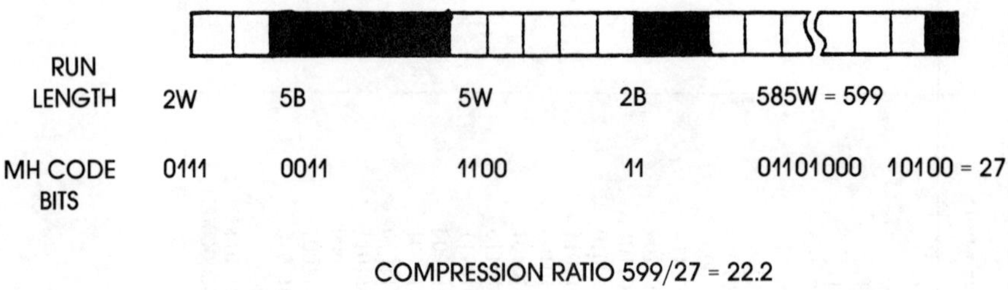

Figure 2.5.1.3 Modified Huffman coding example.

```
   SCAN LINES--
  Reference  wwwwwBBBBwwwwwwBBBBwwwwwwBBwwwwwwwwwwwwBBBBBB
     Coding  wwwwBBBBBwwwwwwwwwwwwwwwwBBwwwwBBBwwwwwBBBBBB
CHANGING PELS--
 Reference Line    *   *     *   *     * *            *
    Coding Line   *    *               * *   *  *     *

        MODE      V    V     P         V V   H        V
                  V=Vertical P=Pass          H=Horizontal

                  w=white pel  B=black pel
```

Figure 2.5.2.1 Changing pels.

	Left				Right		
Coded Pel Position	-3	-2	-1	0	+1	+2	+3
Code	0000010	000010	010	1	001	000011	0000011

Figure 2.5.2.2 Vertical coding of modified read.

If the changing pel pairs are not within ±3 pels, one of two other coding modes is used. The pass mode coding (0001) is used to skip two changing pels on the reference line. Vertical coding is again used for the next pair of changing pels if they are within the ±3-pel vertical range. If two changing pels are on the coding line, the horizontal mode coding (001) is used for modified Huffman coding of the next two runs, and vertical mode will be used again if possible.

When the fax signal is distorted by a noise pulse (lightning, switching transient, or other cause), errors are made in the pattern of black and white dots in the received copy. To prevent this incorrect pattern from propagating down the page, a modified Huffman coded line is sent periodically. At normal resolution, this is done every second line (K = 2), and at fine resolution, every fourth line (K = 4). This is referred to as the *K factor.* (See Figure 2.5.2.3.) Group 4 uses no modified Huffman coded lines (see Chapter 3) (K = ∞).

```
Standard Resolution
 --------------------------------------->     MH coding
 ........................................>     2-D coding
 --------------------------------------->     MH coding
 ........................................>     2-D coding

Fine Resolution
 --------------------------------------->     MH coding
 ........................................>     2-D coding
 ........................................>     2-D coding
 ........................................>     2-D coding
 --------------------------------------->     MH coding
```

Figure 2.5.2.3 K factor for Group 3 fax.

The fastest rate that scanning lines can be sent to a receiving fax unit is determined by the *minimum scan line time* (MSLT), the time taken by the receiving fax unit to print a scan line. The standard MSLT is 20 ms, but it can range from 0 to 40 ms/line depending on the fax equipment design. The fax transmitter obtains this information from the receiver during handshake and never sends faster, but often sends slower. Actual sending time depends on the number of coded bits per line and the modem speed. The number of coded bits is determined by the amount of black and white information. MR coding gives fewer bits than MH coding. The modem speed is set during handshake. If a coded line at the transmitter is ready before the fax receiver is ready, the sending must be delayed by adding fill bits, a string of 0s, which are thrown away at the receiver. No fill bits are used if the fax receiver has full-page memory (0 ms/line).

2.6 MODEM

A modem is a *mo*dulation-*dem*odulation device that accepts the digital information and modulates it into an analog signal to be sent over a voice telephone line. At the receiver, the analog signal is converted back to a duplicate of the digital signal at the transmitter.

The complete modem includes electrical interfaces and control protocols not desirable for fax and would make the system more costly. CCITT Recommendation T.4 specifies modulation and demodulation schemes, with the necessary portions of the CCITT modem built into the Group 3 fax units. Recommendation V.27(ter) is required for the fax data at 4800 and 2400 b/s. Recommendation V.21, channel 2 (return channel) is required for the 300 b/s signaling. Recommendation V.29 is optional for fax data at 9600 and 7200 b/s for use on point-to-point four-wire leased type telephone channels and high-quality switched telephone channels. For convenience, this book will refer to Group 3 modems, meaning the modulation and demodulation, automatic equalization, and other portions of the modems built into Group 3 fax equipment. There is no RS-232 interface with the Group 3 modems.

The Group 3 modems usually try first at 9600 b/s. If training is not successful, they will automatically step down to 7200 b/s and try again. If necessary, this process is repeated at 4800 and 2400 b/s. Most Group 3 fax units have 9600-b/s modems. Fax units with a top rate of 4800 b/s will start at that speed. Higher modem speeds of 14,400 b/s are being considered for fax use on the PSTN. (See Section 2.3.)

Many people incorrectly use bits per second and baud interchangeably. Telephone channels can pass a certain maximum number of changes in signal state per second. This is the *baud rate.* Many slow-speed modems send only one bit of information for each baud. Higher speed modems send more than one bit per baud. Table 2.6.1 shows specifications for the data modems used in Group 3 fax units.

When V.29 is operating at 9600 b/s, it samples the incoming digital fax signal in 4-bit segments. There are 16 different analog signal states at the modem output which represent every possible arrangement of the 0s or 1s of the 4-bit segments at its input. Each of the 16 dots represents a different signal state of amplitude and phase. At 90 degrees, for example, the signal state 1010 has an amplitude of 3 or 0010 has an amplitude of 2. (See Figure 2.6.1.)

Table 2.6.1. Group 3 Modems

Bits per Second	*Baud Rate*	*Bits per Sample*	*Type*	*Carrier Frequency*	*Bandwidth in Hertz*
9600	2400	4	V.29	1700	450–2950
7200	2400	3	V.29	1700	450–2950
4800	1600	3	V.27(ter)	1800	950–2650
2400	1200	2	V.27(ter)	1800	1150–2450

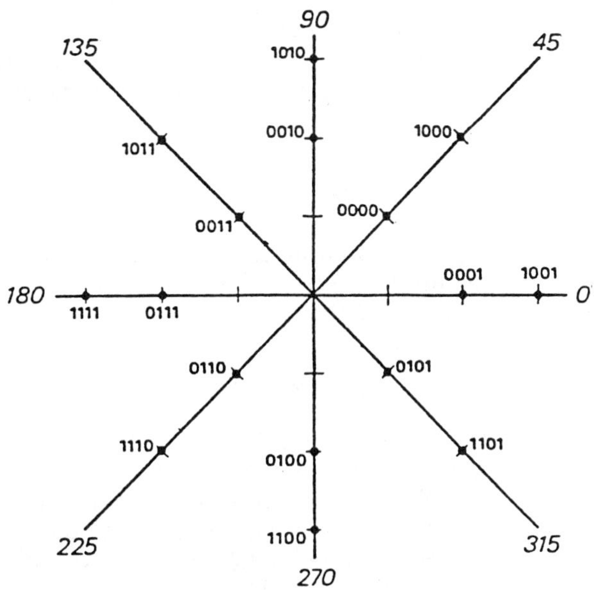

Figure 2.6.1 V.29 signal space diagram at 9600 b/s.

At 9600 b/s, the modem output signal state changes 2400 times per second for a baud rate of 2400. V.29 alternatively can operate at 7200 b/s by sampling 3-bit segments of digital data. This gives eight analog signal states on the telephone line and a 2400-baud rate. Notice that the phone line bandwidth required is the same as for 9600 b/s. Transmission impairments such as phase jitter and noise cause the received signal dots to move around from their assigned positions and be mistaken for another adjacent signal state. This causes errors. The advantage of 7200-b/s operation is a lower error rate on some telephone channels. The modem receiver circuitry makes a signal state decision per baud on only 8 points instead of 16.

The telephone channel bandwidth required for V.29 is 450 to 2950 Hz. Both V.29 and V.27(ter) generate sidebands below and above the carrier equal to the dot-cycle keying rate. This is equal to half the baud rate, giving sidebands at 500 and 2900 Hz for V.29; 50 Hz is added to avoid sharp amplitude or delay curves at the sideband frequencies, making the bandwidth 450 to 2950 Hz. With V.27(ter), the bandwidth is narrower, as shown in Figure 2.6.2.

The envelope-delay distortion from a telephone channel often causes more received signal distortion than the frequency response of the channel. Both of these distortions are minimized by the automatic equalizer after each circuit connection is made. Before sending data, the transmitting modem sends a short training signal of fixed form which is known to the receiver. The receiver tries various combinations of delay, amplitude, and phase of a second "mirror" type of signal that it generates to reverse the telephone line distortion and subtract it. The received bit errors are significantly reduced, allowing most facsimile transmissions to be made to operate at 9600 b/s.

2.7 PROTOCOL

The protocol for Group 3 is given in CCITT Recommendation T.30, Procedures for Document Facsimile Transmission in the General Switched Telephone Network.

2.7.1 Facsimile Call Placement

The telephone number where a Group 3 fax unit is plugged into the phone jack is normally reserved by the user for facsimile traffic instead of voice. In cases where a telephone is not part of the fax unit, a separate telephone is plugged into a phone jack furnished for this purpose in the fax unit. The phone is for dialing other fax units, but it can alternatively be used for voice conversation when there is no fax traffic. The Group 3 fax unit should always be left on and connected to the telephone line so that it can automatically receive fax traffic at any time. The standby power is usually less than that of a 25-watt lamp.

Making a facsimile call and sending a document is divided into five time phases as shown below for a one page transmission:

(1) Phase A — Call Setup.
- Calling fax unit dials the number of the other fax unit.
- Called fax unit automatically answers the ring signal by connecting to line.

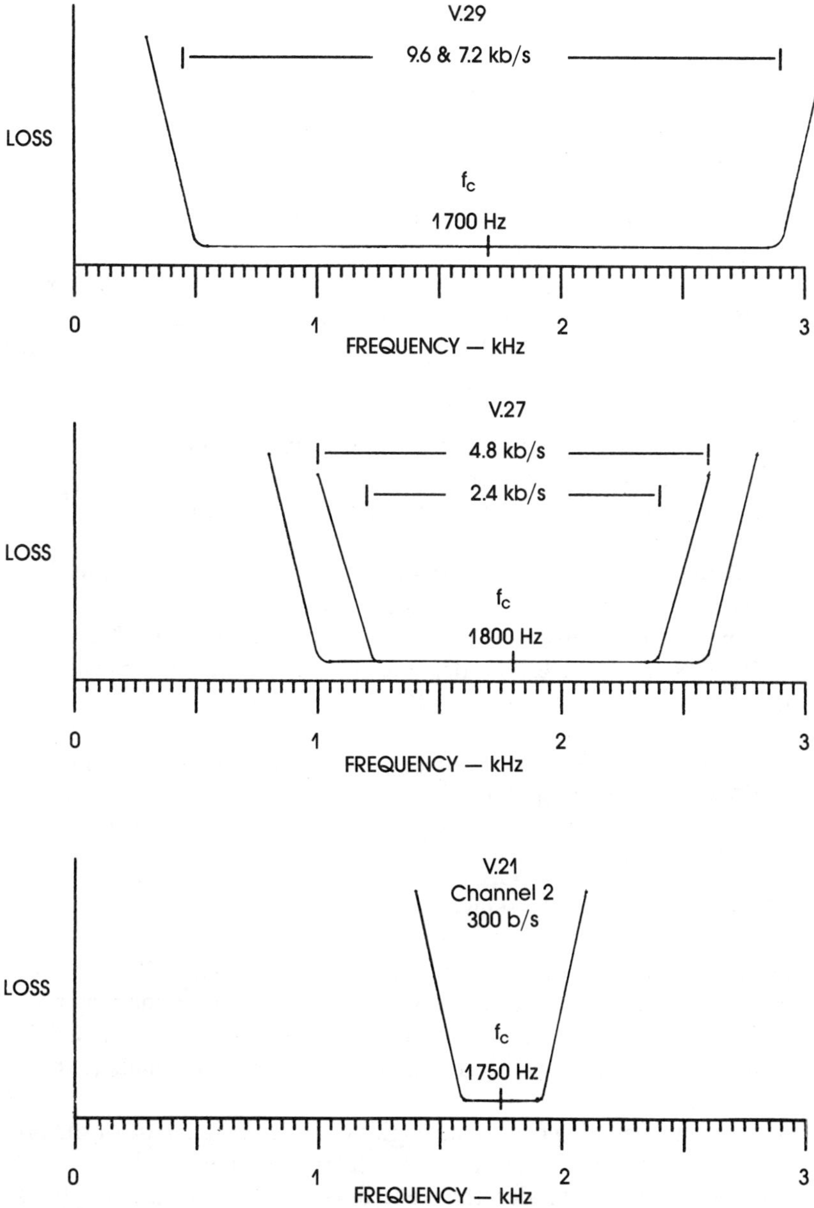

Figure 2.6.2 Group 3 bandwidth requirements.

- Called fax unit sends a 2100 Hz tone back to the calling fax unit for three seconds.

(2) Phase B — Premessage Procedure.

- Called fax unit sends 32- or 40-bit DIS (digital identification signal) at 300 b/s identifying its capabilities, including options of 9600 b/s, 196 lines/inch, MR coding, wide copy, minimum-recording-line-time, and error-correcting mode.
- Calling fax unit sends DCS (digital command signal), locking in the called unit to the capabilities selected.
- Calling fax unit sends high-speed long training signal for the data modem.
- Called fax unit sends CFR (confirmation to receive) signal to confirm the receiving modem is trained (adjusted for low-error operation) and that the fax unit is ready to receive.

(3) Phase C — Fax Message Transmission

- Calling fax unit sends short training signal and picture signals.

(4) Phase D — Postmessage Procedure

- At the end of the page being sent, the calling fax unit sends EOP (end of procedure) signal.
- Called fax unit sends MCF (message confirmation) to indicate the page was received with errors within acceptable limits.

(The maximum acceptable errors that can be received before returning a negative response (PIP, PIN, RTP, or RTN), depends on the fax receiver internal setting. Settings up to 32 or 64 lines per page with errors are accepted.)

(5) Phase E — Call Release

- Calling fax unit sends DCN (disconnect) signal and both fax units disconnect from the telephone line.

2.7.2 Signaling for Calling Party Sends

Many fax units are programmed to answer an incoming call at the start of the first ring, so the ring on a telephone connected to the same line may not even be heard. After answering the ring, the called fax unit sends a 2100 Hz CED tone for three seconds, followed by a short burst of 300 b/s DIS signal. (See Figure 2.7.2.1.)

When the person making the call hears this distinctive signal, he or she presses the send button to connect the fax unit to the telephone line.

The calling fax unit decodes the next burst of DIS, which identifies the capabilities of the called fax unit. (See Table 2.7.2.1.)

The calling fax unit sends a burst of DCS signal to select the compatible features to be used for this call. Appended to DCS is a short training signal for

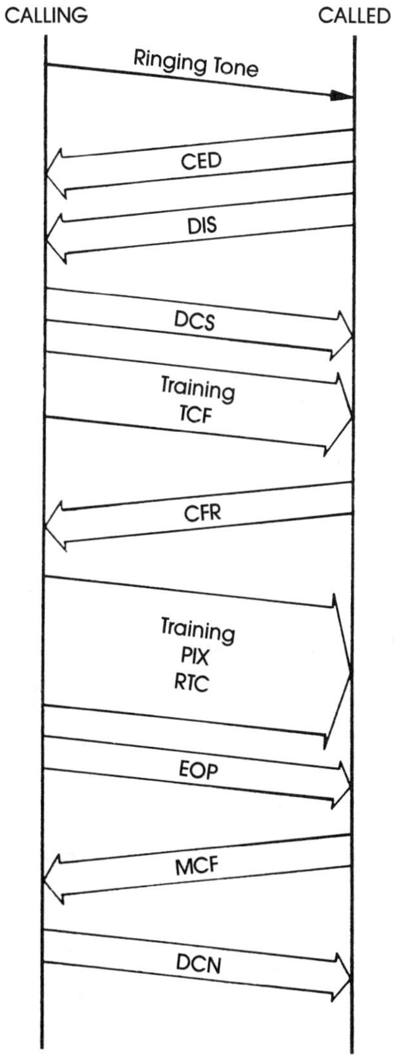

Figure 2.7.2.1 Signaling for calling party sends.

the receiving modem and a 1.5-second TCF signal (all 0s) to check whether the modem is receiving error-free bits. If these signals are received correctly, CFR signal is sent from the receiver. The calling station then sends a short modem training signal followed by the fax signals. At the end of fax signals for one page, RTC command is sent to switch back to the 300-b/s modem, followed by EOP

Table 2.7.2.1. Handshake Abbreviations

Abbreviation	*Function*	*Signal Format*
CED	Called station identification	2100 Hz
CFR	Confirmation to receive	X010 0001 1850 or 1650 Hz for 3 s
CRP	Command repeat	X101 1000
CIG	Calling subscriber identification	1000 0010
CNG	Calling tone	1100 Hz for 500 ms
CSI	Called subscriber identification	0000 0010
CTC	Continue to correct	X100 1000
CTR	Response to continue to correct	X010 0011
DCN	Disconnect	X101 1111
DCS	Digital command signal	X100 0001
DIS	Digital identification signal	0000 0001
DTC	Digital transmit command	1000 0001
EOM	End of message	X111 0001 1100 Hz
EOP	End of procedure	X111 0100
EOR	End of retransmission	X111 0011
ERR	Response to end of retransmission	X011 1000
FCD	Facsimile coded data	0110 0000
FCF	Facsimile control field	
FIF	Facsimile information field	
FIT	Failure to train	X010 0010
GC	Group command	1300 Hz for 1.5–10.0 s 2100 Hz for 1.5–10.0 s

Table 2.7.2.1. Handshake Abbreviations (Continued.)

Abbreviation	*Function*	*Signal Format*
GI	Group identification	1650 or 1850 Hz
HDLC	High level data link control	
LCS	Line conditioning signals	1100 Hz
MCF	Message confirmation	X011 0001 1650 or 1850 Hz
MPS	Multipage signal	X111 0010
NSC	Nonstandard facilities command	1000 0100
NSF	Nonstandard facilities	0000 0100
NSS	Nonstandard setup	X100 0100
PIN	Procedural interrupt negative	X011 0100
PIP	Procedural interrupt positive	X011 0101
PIS	Procedure interrupt signal	462 Hz for 3 s
PPS	Partial page signal	X111 1101
PPR	Partial page request	X011 1101
PRI-EOM	Procedure interrupt-EOM	X111 1001
PRI-EOP	Procedure interrupt-EOP	X111 1100
PRI-MPS	Procedure interrupt-MPS	X111 1010
RCP	Return to control for partial page	0110 0001
RNR	Receive not ready	X011 0111
RR	Receive ready	X111 0110
RTN	Retrain negative	X011 0010
RTP	Retrain positive	X011 0011
TCF	Training check	Zeros for 1.5 s
TSI	Transmitting subscriber identification	X100 0010

(end of [fax] procedure). The fax receiver sends MCF to indicate the page has been received correctly. The fax transmitter notes "OK" in its log and sends DCN. Both fax units disconnect from the telephone line.

2.7.3 Optional and Nonstandard Facilities Call

Valuable optional proprietary features can be used within the protocol for Group 3 fax units. These features are invoked during handshake in a manner that will not interfere with the basic Group 3 service. (See Figure 2.7.3.1)

NSF (nonstandard facilities) and CSI (called subscriber identification) are added ahead of DIS. NSF means the fax receiver has the ability to use proprietary features not covered by the Series-T recommendations. After receiving NSF, the calling fax unit returns NSS (nonstandard setup) to lock in the called fax unit to the nonstandard operation specified. Many manufacturers have a private error-correction system in use between their own fax units. It is mandatory that Group 3 fax units with NSS capabilities also have standard capabilities to operate with other standard fax units.

CSI sends a coded signal with the telephone number of the called fax unit. Some fax units will display this number so that the fax operator can check that he or she is sending to the correct unit, and then the sending fax records the number in its electronic log. The sending fax uses TSI (transmit subscriber identification) to send its telephone number to the receiving fax.

2.7.4 Polling Called Party

Polling allows an orderly sending of once-a-day information from many fax units to a single fax at a central point. The central fax can be programmed to call the sending units sequentially and command each one to send documents from memory or its automatic document feeder. Without polling it would be very difficult to prevent line busy conditions at the central point. Many Group 3 fax units have delayed calling programs with an automatic dialer which will allow polling at night when there are no operators around. (See Figure 2.7.4.1 for polling protocol.)

After the central calling fax unit receives the called fax DIS response, it sends NSC (nonstandard command) and the other fax unit takes over the session until it has sent all of the pages destined for the calling fax unit. CIG (calling subscriber identification) is sent instead of TSS, with the telephone number coded in the same format. The subscriber identification codes may be checked by the transmitter to ensure that the documents are being sent to the right caller.

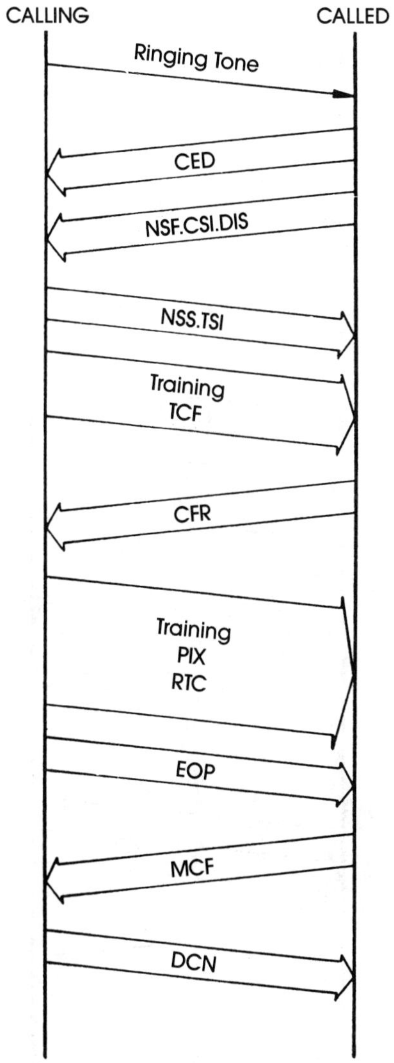

Figure 2.7.3.1 Nonstandard facilities call.

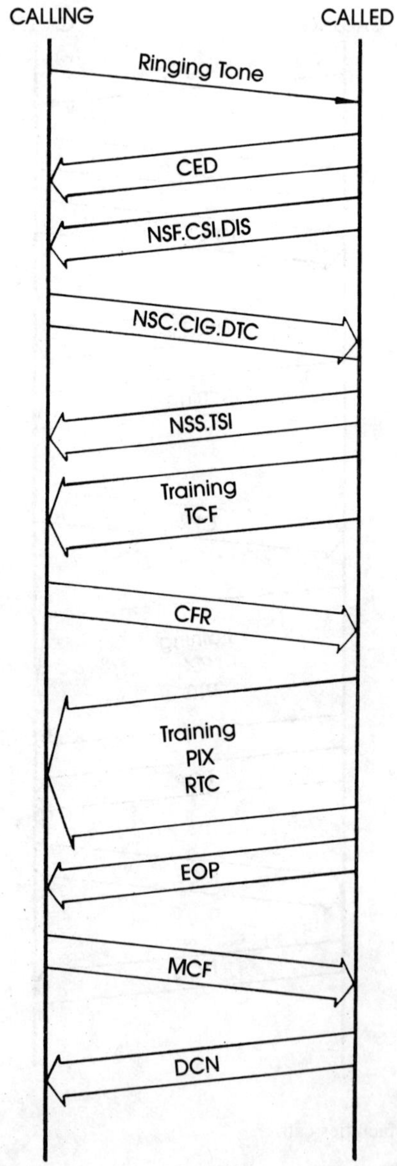

Figure 2.7.4.1 Polling called party.

At the start of the initial handshake, the receiver uses its 300-b/s modem to send back information to the transmitter. Because this modem does not need a training signal to operate with a low error rate, the handshaking procedure is simplified. Optional handshake at 2400 b/s is possible, but seldom used. High level data link control (HDLC) frames are used for the binary coded handshaking.

The preamble signal of flag fields (0111 1110) for one second precedes the actual DIS data making sure the telephone line echo suppressors are operating in the right direction. Frame synchronization is also derived from this signal. The HDLC *address field,* the next byte, is 1111 1111, and it is followed by the HDLC *control field* which is 1100 1000 for the last frame sent prior to an expected response from the distant station, or 1100 0000 for other frames. The frame error-checking sequence is a two-byte signal calculated from the content of the data being sent. The receiving unit calculates the same digital number to confirm validity of the handshake signal data. (See Table 2.7.2.1.)

2.7.6 DIS Signal

The DIS is a 300-b/s handshake signal sent when the fax unit answers a call. The optional NSF and CSI frames, if used, must be sent ahead of the standard Group 3 capabilities information. Forty bits are currently used under the Blue Book recommendations as shown in Table 2.7.6.1.

Older Group 3 fax units that still send 32 bits do not have the extra features, but are compatible. Additions have been made from the 32-bit Red Book version to add the error-correction mode and the smaller page size A5 and A6 Group 3 fax units. NSF (0000 0100) is used to invoke nonstandard features between two fax units made by the same manufacturer. This must be followed immediately by one byte of country code (CCITT Recommendation T.35) and at least one more byte coded in any way desired to describe the nonstandard features. The features specified by NSF are not recognized by fax units made by other manufacturers. If handshake is completed in NSF mode, the format for fax transmission could be changed to achieve higher speeds or for other purposes. The NSF method allows fax system designs to be improved without affecting compatibility with other Group 3 units.

CSI, CIG, and TSI are each one byte, followed by one byte for a + and one byte for each digit of the international telephone number of the fax unit. The telephone number should be programmed into the fax unit after it is installed on the telephone line, and reprogrammed if moved to another telephone number.

Table 2.7.6.1. DIS and DTC Bit Settings

Bit Number	*DIS and DTC*	*DCS*
1	Transmitter – T.2 operation	
2	Receiver – T.2 operation	Receiver – T.2 operation
3	T.2 IOC = 176	T.2 IOC = 176
4	Transmitter – T.3 operation	
5	Receiver – T.3 operation	Receiver – T.3 operation
6	Reserved for future T.3 operation features	
7	Reserved for future T.3 operation features	
8	Reserved for future T.3 operation features	
9	Transmitter – T.4 operation	
10	Receiver – T.4 operation	Receiver – T.4 operation
11, 12	Data signaling rate:	Data signaling rate:
(0,0)	V.27 (ter) fallback mode	2400 bits/s V.27 (ter)
(0,1)	V.27 (ter)	4800 bits/s V.27 (ter)
(1,0)	V.29	9600 bits/s V.29
(1,1)	V.27 (ter) and V.29	7200 bits/s V.29
13	Reserved for new modulation system	
14	Reserved for new modulation system	
15	Vertical resolution = 7.7 lines/mm	Vertical resolution = 7.7 lines/mm
16	Two-dimensional coding capability	Two-dimensional coding capability
17, 18	Recording width capabilities:	Recording width:
(0,0)	1728 picture elements along scan line length of 215 mm ± 1%	1728 picture elements along scan line length of 215 mm ± 1%
(0,1)	1728 picture elements along scan line length of 215 mm ± 1% and 2048 picture elements along scan line length of 255 mm ± 1% and 2432 picture elements along scan line length of 303 mm ± 1%	2432 picture elements along scan line length of 303 mm ± 1%
(1,0)	1728 picture elements along scan line length of 215 mm ± 1% and 2048 picture elements along scan line length of 255 mm ± 1%	2048 picture elements along scan line length of 255 mm ± 1%
(1,1)	Invalid (*See* Note 7.)	Invalid (*See* Note 7.)
19, 20	Maximum recording length capability:	Maximum recording length capability:
(0,0)	A4 (297 mm)	A4 (297 mm)
(0,1)	Unlimited	Unlimited
(1,0)	A4 (297 mm) and B4 (364 mm)	B4 (364 mm)
(1,1)	Invalid	Invalid

Table 2.7.6.1. DIS and DTC Bit Settings (Continued.)

Bit Number	DIS and DTC	DCS
21, 22, 23	Minimum scan line time capability at the receiver:	Minimum scan line time:
(0,0,0)	20 ms at 3.851/mm: T7.7-T3.85	20 ms
(0,0,1)	40 ms at 3.851/mm: T7.7-T3.85	40 ms
(0,1,0)	10 ms at 3.851/mm: T7.7-T3.85	10 ms
(1,0,0)	5 ms at 3.851/mm: T7.7-T3.85	5 ms
(0,1,1)	10 ms at 3.851/mm: T7.7-1/2 T3.85	
(1,1,0)	20 ms at 3.851/mm: T7.7-1/2 T3.85	
(1,0,1)	40 ms at 3.851/mm: T7.7-1/2 T3.85	
(1,1,1)	0 ms at 3.851/mm: T7.7-T3.85	0 ms
24	Extend field	Extend field
25	2400 bits/s handshaking	2400 bits/s handshaking
26	Uncompressed mode	Uncompressed mode
27	Error correction mode	Error correction mode
28	Set to "0"	Frame size: 0 = 256 octets 1 = 64 octets
29	Error limiting mode	Error limiting mode
30	Reserved for Group 4 capability on PSTN	Reserved for Group 4 capability on PSTN
31	Unassigned	Unassigned
32	Extend field	Extend field
33	Validity of bits 17, 18	Recording width:
(0)	Bits 17, 18 are valid	Recording width indicated by bits 17, 18
(1)	Bits 17, 18 are invalid	Recording bits indicated by this field bit information
34	Recording width capability: 1216 picture elements along scan line length of 151 mm ± 1%	Middle 1216 elements of 1728 picture elements
35	Recording width capability: 864 picture elements along scan line length of 107 mm ± 1%	Middle 864 elements of 1728 picture elements
36	Recording width capability: 1728 picture elements along scan line length of 151 mm ± 1%	Invalid
37	Recording width capability: 1728 picture elements along scan line length of 107 mm ± 1%	Invalid
38	Reserved for future recording width capability	
39	Reserved for future recording width capability	
40	Extend field	Extend field

The other fax unit, upon receiving this number, may display it, record it in its log, or print it on received copy. Another method sometimes used for subscriber identification is internally generating a fax signal with the desired information and adding it to the page.

2.8 GROUP 3 ENHANCEMENTS

Flexibility was provided by incorporating unassigned bits in the Group 3 protocol. Some of these bits were assigned years after the standard was adopted (e.g., error-correction mode).

2.8.1 Error Concealment Techniques

The switched telephone network is prone to error when transmitting digital data. When such a transmission error occurs, a streak is typically created on the output page which can be disturbing to the observer. The purpose of error concealment techniques is to minimize the visual effect of transmission errors on the output copy. Error concealment is possible since an end-of-line code is transmitted between scan lines. If the receiver does not detect the EOL code at the expected location in the data stream, an error has likely occurred on the line following the last correctly received EOL code. Some examples of possible error conditions are:

- EOL occurs before 1728 pels have been written;
- More than 1728 pels have been written before EOL is received;
- No word in applicable code tables matches received bit pattern;
- Current line decoding references a run that does not exist in the previous line.

The effect of errors on output copy has been quantitatively measured by means of computer simulation; the simulation process is illustrated in Figure 2.8.1.1. To assist in interpreting these results, a new term, the *error sensitivity factor* (ESF), is used. The ESF is defined as the number of incorrect pels in the output image, resulting from transmission errors, divided by the number of transmitted bits in error. In other words, the ESF represents the effect of a typical transmission bit error on the picture.

Simulation runs were performed on the *modified Huffman code* (MH) and the *modified read code* (MR) using the parameters shown in Table 2.8.1.1. As indicated in the table, the MR is twice as sensitive to errors as the MH even though an error concealment technique was employed on the MR. This increased sensitivity is expected because the MR achieves greater compression by reducing redundancy in two dimensions. Figures 2.8.1.2 and 2.8.1.3 illustrate the output images from this simulation. Note that the transmission errors have had a greater

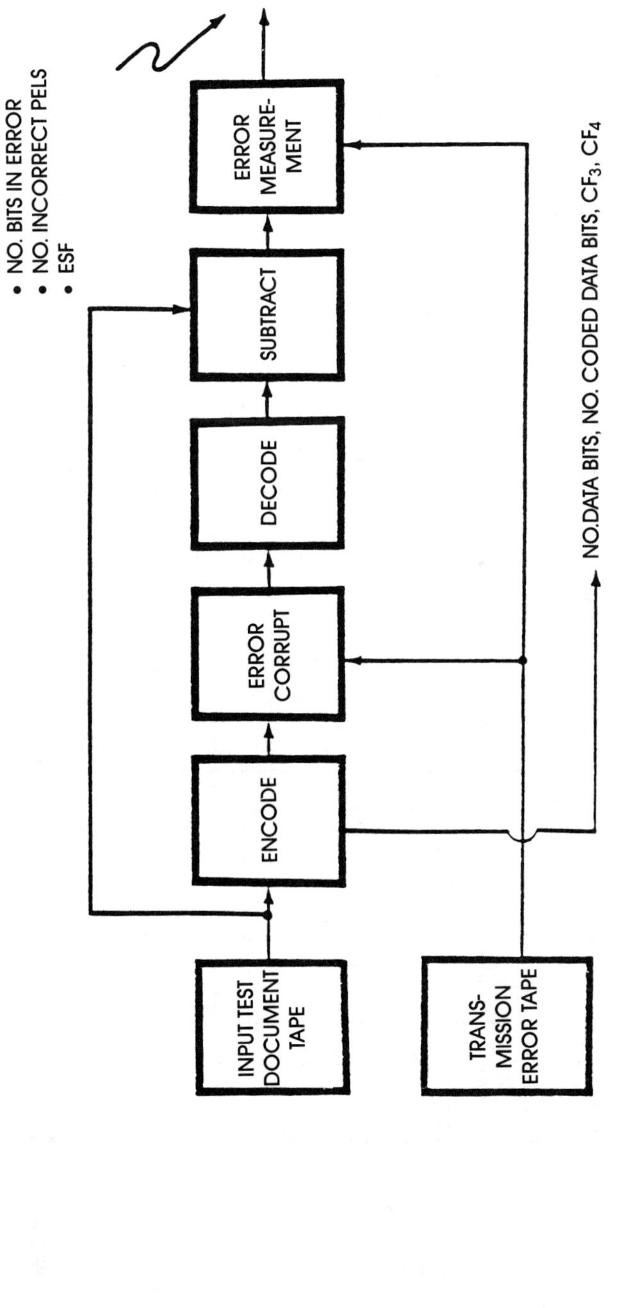

Figure 2.8.1.1 Block diagram of simulation process.

Table 2.8.1.1 Comparison of modified Huffman and modified read codes.

TEST PARAMETERS				MODIFIED HUFFMAN		MODIFIED READ		
TEST DOC	VERT RESOL.	MSLT	ERROR FILE	BER $\times 10^{-3}$	ESF	K	BER $\times 10^{-3}$	ESF
4	3.85	10	1	.82	24.93	2	.66	40.31
4	3.85	20	1	.81	27.64	2	.82	45.24
4	7.7	20	1	.65	29.28	4	.78	58.70
4	3.85	20	2	1.1	19.23	2	1.16	27.69
4	3.85	20	3	.70	20.98	2	.67	57.41
4	3.85	20	4	1.3	14.66	2	1.36	21.22
1	3.85	20	1	.81	9.70	2	.63	20.60
7	3.85	20	1	.82	22.06	2	.69	30.27
Average ESF					21.06			37.68

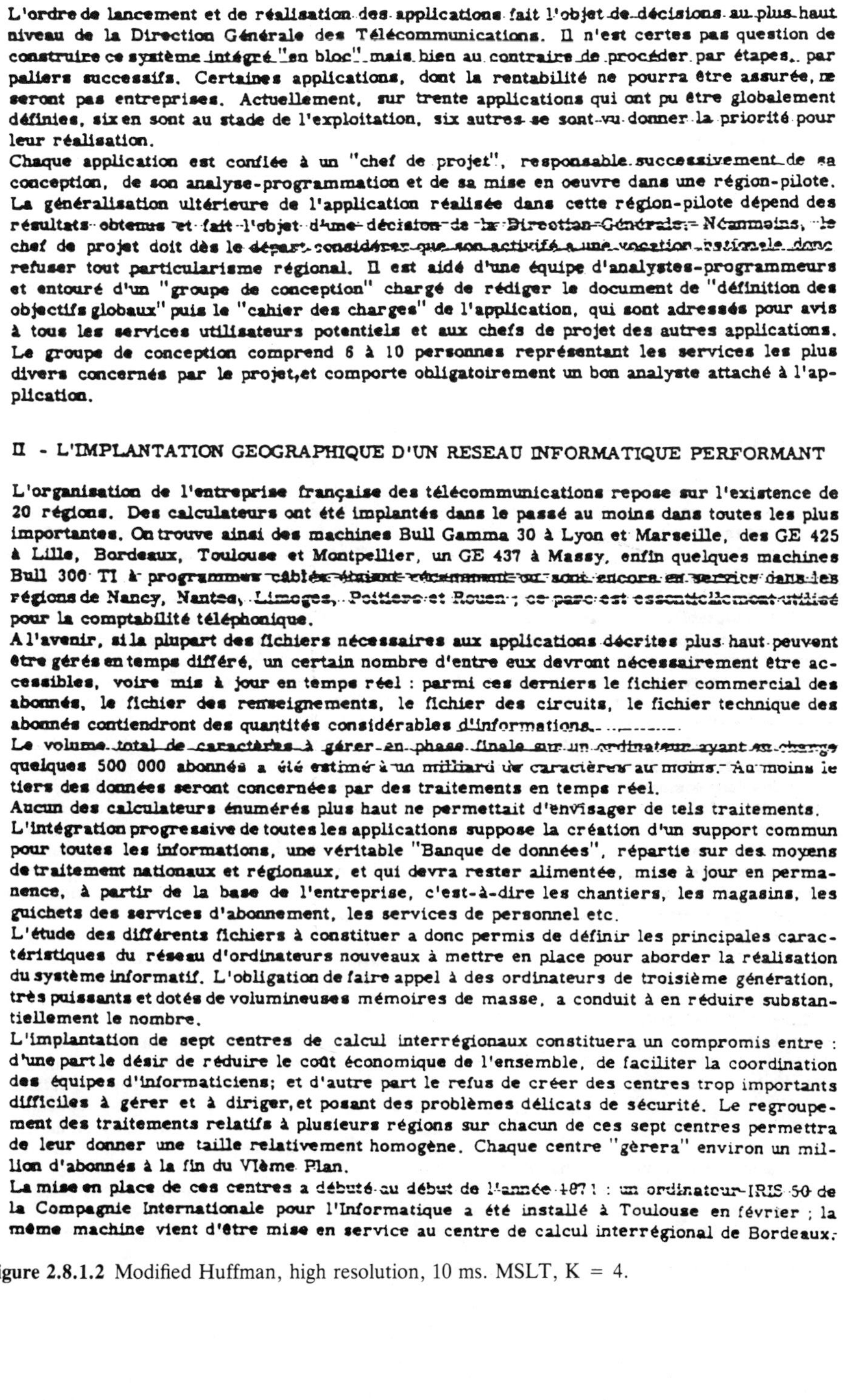

L'ordre de lancement et de réalisation des applications fait l'objet de décisions au plus haut niveau de la Direction Générale des Télécommunications. Il n'est certes pas question de construire ce système intégré "en bloc" mais bien au contraire de procéder par étapes, par paliers successifs. Certaines applications, dont la rentabilité ne pourra être assurée, ne seront pas entreprises. Actuellement, sur trente applications qui ont pu être globalement définies, six en sont au stade de l'exploitation, six autres se sont vu donner la priorité pour leur réalisation.
Chaque application est confiée à un "chef de projet", responsable successivement de sa conception, de son analyse-programmation et de sa mise en oeuvre dans une région-pilote. La généralisation ultérieure de l'application réalisée dans cette région-pilote dépend des résultats obtenus et fait l'objet d'une décision de la Direction Générale. Néanmoins, le chef de projet doit dès le départ considérer que son activité a une vocation nationale donc refuser tout particularisme régional. Il est aidé d'une équipe d'analystes-programmeurs et entouré d'un "groupe de conception" chargé de rédiger le document de "définition des objectifs globaux" puis le "cahier des charges" de l'application, qui sont adressés pour avis à tous les services utilisateurs potentiels et aux chefs de projet des autres applications. Le groupe de conception comprend 6 à 10 personnes représentant les services les plus divers concernés par le projet, et comporte obligatoirement un bon analyste attaché à l'application.

II - L'IMPLANTATION GEOGRAPHIQUE D'UN RESEAU INFORMATIQUE PERFORMANT

L'organisation de l'entreprise française des télécommunications repose sur l'existence de 20 régions. Des calculateurs ont été implantés dans le passé au moins dans toutes les plus importantes. On trouve ainsi des machines Bull Gamma 30 à Lyon et Marseille, des GE 425 à Lille, Bordeaux, Toulouse et Montpellier, un GE 437 à Massy, enfin quelques machines Bull 300 TI à programmes câblés étaient récemment ou sont encore en service dans les régions de Nancy, Nantes, Limoges, Poitiers et Rouen ; ce parc est essentiellement utilisé pour la comptabilité téléphonique.
A l'avenir, si la plupart des fichiers nécessaires aux applications décrites plus haut peuvent être gérés en temps différé, un certain nombre d'entre eux devront nécessairement être accessibles, voire mis à jour en temps réel : parmi ces derniers le fichier commercial des abonnés, le fichier des renseignements, le fichier des circuits, le fichier technique des abonnés contiendront des quantités considérables d'informations.
Le volume total de caractères à gérer en phase finale sur un ordinateur ayant en charge quelques 500 000 abonnés a été estimé à un milliard de caractères au moins. Au moins le tiers des données seront concernées par des traitements en temps réel.
Aucun des calculateurs énumérés plus haut ne permettait d'envisager de tels traitements.
L'intégration progressive de toutes les applications suppose la création d'un support commun pour toutes les informations, une véritable "Banque de données", répartie sur des moyens de traitement nationaux et régionaux, et qui devra rester alimentée, mise à jour en permanence, à partir de la base de l'entreprise, c'est-à-dire les chantiers, les magasins, les guichets des services d'abonnement, les services de personnel etc.
L'étude des différents fichiers à constituer a donc permis de définir les principales caractéristiques du réseau d'ordinateurs nouveaux à mettre en place pour aborder la réalisation du système informatif. L'obligation de faire appel à des ordinateurs de troisième génération, très puissants et dotés de volumineuses mémoires de masse, a conduit à en réduire substantiellement le nombre.
L'implantation de sept centres de calcul interrégionaux constituera un compromis entre : d'une part le désir de réduire le coût économique de l'ensemble, de faciliter la coordination des équipes d'informaticiens; et d'autre part le refus de créer des centres trop importants difficiles à gérer et à diriger, et posant des problèmes délicats de sécurité. Le regroupement des traitements relatifs à plusieurs régions sur chacun de ces sept centres permettra de leur donner une taille relativement homogène. Chaque centre "gèrera" environ un million d'abonnés à la fin du VIème Plan.
La mise en place de ces centres a débuté au début de l'année 1971 : un ordinateur IRIS 50 de la Compagnie Internationale pour l'Informatique a été installé à Toulouse en février ; la même machine vient d'être mise en service au centre de calcul interrégional de Bordeaux.

Figure 2.8.1.2 Modified Huffman, high resolution, 10 ms. MSLT, K = 4.

L'ordre de lancement et de réalisation des applications fait l'objet de décisions au plus haut niveau de la Direction Générale des Télécommunications. Il n'est certes pas question de construire ce système intégré "en bloc" mais bien au contraire de procéder par étapes, par paliers successifs. Certaines applications, dont la rentabilité ne pourra être assurée, ne seront pas entreprises. Actuellement, sur trente applications qui ont pu être globalement définies, six en sont au stade de l'exploitation, six autres se sont vu donner la priorité pour leur réalisation.
Chaque application est confiée à un "chef de projet", responsable successivement de sa conception, de son analyse-programmation et de sa mise en oeuvre dans une région-pilote. La généralisation ultérieure de l'application réalisée dans cette région-pilote dépend des résultats obtenus et fait l'objet d'une décision de la Direction Générale. Néanmoins, le chef de projet doit dès le départ considérer que son activité a une vocation nationale donc refuser tout particularisme régional. Il est aidé d'une équipe d'analystes-programmeurs et entouré d'un "groupe de conception" chargé de rédiger le document de "définition des objectifs globaux" puis le "cahier des charges" de l'application, qui sont adressés pour avis à tous les services utilisateurs potentiels et aux chefs de projet des autres applications. Le groupe de conception comprend 8 à 10 personnes représentant les services les plus divers concernés par le projet, et comporte obligatoirement un bon analyste attaché à l'application.

II - L'IMPLANTATION GEOGRAPHIQUE D'UN RESEAU INFORMATIQUE PERFORMANT

L'organisation de l'entreprise française des télécommunications repose sur l'existence de 20 régions. Des calculateurs ont été implantés dans le passé au moins dans toutes les plus importantes. On trouve ainsi des machines Bull Gamma 30 à Lyon et Marseille, des GE 425 à Lille, Bordeaux, Toulouse et Montpellier, un GE 437 à Massy, enfin quelques machines Bull 300 TI à programmes câblés étaient récemment ou sont encore en service dans les régions de Nancy, Nantes, Limoges, Poitiers et Rouen ; ce parc est essentiellement utilisé pour la comptabilité téléphonique.
A l'avenir, si la plupart des fichiers nécessaires aux applications décrites plus haut peuvent être gérés en temps différé, un certain nombre d'entre eux devront nécessairement être accessibles, voire mis à jour en temps réel : parmi ces derniers le fichier commercial des abonnés, le fichier des renseignements, le fichier des circuits, le fichier technique des abonnés contiendront des quantités considérables d'informations.
Le volume total de caractères à gérer en phase finale sur un ordinateur ayant en charge quelques 500 000 abonnés a été estimé à un milliard de caractères au moins. Au moins le tiers des données seront concernées par des traitements en temps réel.
Aucun des calculateurs énumérés plus haut ne permettait d'envisager de tels traitements.
L'intégration progressive de toutes les applications suppose la création d'un support commun pour toutes les informations, une véritable "Banque de données", répartie sur des moyens de traitement nationaux et régionaux, et qui devra rester alimentée, mise à jour en permanence, à partir de la base de l'entreprise, c'est-à-dire les chantiers, les magasins, les guichets des services d'abonnement, les services de personnel etc.
L'étude des différents fichiers à constituer a donc permis de définir les principales caractéristiques du réseau d'ordinateurs nouveaux à mettre en place pour aborder la réalisation du système informatif. L'obligation de faire appel à des ordinateurs de troisième génération, très puissants et dotés de volumineuses mémoires de masse, a conduit à en réduire substantiellement le nombre.
L'implantation de sept centres de calcul interrégionaux constituera un compromis entre : d'une part le désir de réduire le coût économique de l'ensemble, de faciliter la coordination des équipes d'informaticiens; et d'autre part le refus de créer des centres trop importants difficiles à gérer et à diriger, et posant des problèmes délicats de sécurité. Le regroupement des traitements relatifs à plusieurs régions sur chacun de ces sept centres permettra de leur donner une taille relativement homogène. Chaque centre "gèrera" environ un million d'abonnés à la fin du VIème Plan.
La mise en place de ces centres a débuté au début de l'année 1971 : un ordinateur IRIS 50 de la Compagnie Internationale pour l'Informatique a été installé à Toulouse en février ; la même machine vient d'être mise en service au centre de calcul interrégional de Bordeaux.

Figure 2.8.1.3 Modified read, high resolution, 10 ms. MSLT, K = 4.

subjective effect on the MR image than the MH image. This is consistent with the ESF results.

Simulation runs were performed to compare the four different error concealment techniques for the MR which are described below.

- *Print White (PW)* — The first erroneous line is printed white, and all subsequent lines are printed white until a one-dimensional MH line is correctly received.
- *Print Previous Line (PPL)* — The first erroneous line (x) is replaced by the previous correctly received line ($x - 1$), and all subsequent lines are replaced by $x - 1$ until a one-dimensional MH line is correctly received.
- *Print Previous Line/White (PLW)* — This processing technique is a combination of the previous two. The first erroneous line (x) is replaced by the previous correctly received line ($x - 1$), and all subsequent lines are printed white until a one-dimensional MH line is correctly received.
- *Normal Decode/Previous Line (NDPL)* — In this case the first erroneous line is decoded and printed in the normal MH or MR manner up to the point in the line where the error is detected. From this point, the remainder of the first erroneous line is replaced by the corresponding pels in the "previous line." The resultant "correct" line is then used as a new reference "previous line," and the process is repeated until an MH line is correctly decoded. This error processing technique should be particularly advantageous in those instances where a transmission error occurs near the end (right-hand side of the page) of a scan line period. When this occurs it should be possible to decode correctly most of the scan line that was "hit" as well as most of the subsequent scan lines before a correct MH line is received.

Figures 2.8.1.4, 2.8.1.5, 2.8.1.6, and 2.8.1.7 illustrate the output images respectively resulting from these simulations.

2.8.2 Group 3 Error Control Option

Conventional facsimile equipment employs no error control technique in the classic sense. The periodic transmission of a MH line in the optional MR mode of operation is a type of forward error correction to prohibit the propagation of an error throughout the entire page. Another type of error control is that some fax machines will compute the number of lines in error at the receiver by using the redundant EOL code, and request the retransmission of a page if the number of lines in error exceeds a threshold.

It has long been thought that a more conventional error control technique would be a useful option for Group 3 fax. If such an option were employed in the MR mode, the K factor could be infinity, thereby reducing the transmission time

L'ordre de lancement et de réalisation des applications fait l'objet de décisions au plus haut niveau de la Direction Générale des Télécommunications. Il n'est certes pas question de construire ce système intégré "en bloc" mais bien au contraire de procéder par étapes, par paliers successifs. Certaines applications, dont la rentabilité ne pourra être assurée, ne seront pas entreprises. Actuellement, sur trente applications qui ont pu être globalement définies, six en sont au stade de l'exploitation, six autres se sont vu donner la priorité pour leur réalisation.

Chaque application est confiée à un "chef de projet", responsable successivement de sa conception, de son analyse-programmation et de sa mise en oeuvre dans une région-pilote. La généralisation ultérieure de l'application réalisée dans cette région-pilote dépend des résultats obtenus et fait l'objet d'une décision de la Direction Générale. Néanmoins, le chef de projet doit dès le départ considérer que son activité a une vocation nationale donc refuser tout particularisme régional. Il est aidé d'une équipe d'analystes-programmeurs et entouré d'un "groupe de conception" chargé de rédiger le document de "définition des objectifs globaux" puis le "cahier des charges" de l'application qui sont adressés pour avis à tous les services utilisateurs potentiels et aux chefs de projet des autres applications. Le groupe de conception comprend 6 à 10 personnes représentant les services les plus divers concernés par le projet, et comporte obligatoirement un bon analyste attaché à l'application.

II - L'IMPLANTATION GEOGRAPHIQUE D'UN RESEAU INFORMATIQUE PERFORMANT

L'organisation de l'entreprise française des télécommunications repose sur l'existence de 20 régions. Des calculateurs ont été implantés dans le passé au moins dans toutes les plus importantes. On trouve ainsi des machines Bull Gamma 30 à Lyon et Marseille, des GE 425 à Lille, Bordeaux, Toulouse et Montpellier, un GE 437 à Massy, enfin quelques machines Bull 300 TI à programmes câblés étaient récemment ou sont encore en service dans les régions de Nancy, Nantes, Limoges, Poitiers et Rouen ; ce parc est essentiellement utilisé pour la comptabilité téléphonique.

A l'avenir, si la plupart des fichiers nécessaires aux applications décrites plus haut peuvent être gérés en temps différé, un certain nombre d'entre eux devront nécessairement être accessibles, voire mis à jour en temps réel : parmi ces derniers le fichier commercial des abonnés, le fichier des renseignements, le fichier des circuits, le fichier technique des abonnés contiendront des quantités considérables d'informations.

Le volume total de caractères à gérer en phase finale sur un ordinateur ayant en charge quelques 500 000 abonnés a été estimé à un milliard de caractères au moins. Au moins le tiers des données seront concernées par des traitements en temps réel.

Aucun des calculateurs énumérés plus haut ne permettait d'envisager de tels traitements.

L'intégration progressive de toutes les applications suppose la création d'un support commun pour toutes les informations, une véritable "Banque de données", répartie sur des moyens de traitement nationaux et régionaux, et qui devra rester alimentée, mise à jour en permanence, à partir de la base de l'entreprise, c'est-à-dire les chantiers, les magasins, les guichets des services d'abonnement, les services de personnel etc.

L'étude des différents fichiers à constituer a donc permis de définir les principales caractéristiques du réseau d'ordinateurs nouveaux à mettre en place pour aborder la réalisation du système informatif. L'obligation de faire appel à des ordinateurs de troisième génération, très puissants et dotés de volumineuses mémoires de masse, a conduit à en réduire substantiellement le nombre.

L'implantation de sept centres de calcul interrégionaux constituera un compromis entre : d'une part le désir de réduire le coût économique de l'ensemble, de faciliter la coordination des équipes d'informaticiens; et d'autre part le refus de créer des centres trop importants difficiles à gérer et à diriger, et posant des problèmes délicats de sécurité. Le regroupement des traitements relatifs à plusieurs régions sur chacun de ces sept centres permettra de leur donner une taille relativement homogène. Chaque centre "gèrera" environ un million d'abonnés à la fin du VIème Plan.

La mise en place de ces centres a débuté au début de l'année 1971 : un ordinateur IRIS 50 de la Compagnie Internationale pour l'Informatique a été installé à Toulouse en février ; la même machine vient d'être mise en service au centre de calcul interrégional de Bordeaux.

Figure 2.8.1.4 Print white (7.7 lines/mm).

L'ordre de lancement et de réalisation des applications fait l'objet de décisions au plus haut niveau de la Direction Générale des Télécommunications. Il n'est certes pas question de construire ce système intégré "en bloc" mais bien au contraire de procéder par étapes, par paliers successifs. Certaines applications, dont la rentabilité ne pourra être assurée, ne seront pas entreprises. Actuellement, sur trente applications qui ont pu être globalement définies, six en sont au stade de l'exploitation, six autres se sont vu donner la priorité pour leur réalisation.
Chaque application est confiée à un "chef de projet", responsable successivement de sa conception, de son analyse-programmation et de sa mise en oeuvre dans une région-pilote. La généralisation ultérieure de l'application réalisée dans cette région-pilote dépend des résultats obtenus et fait l'objet d'une décision de la Direction Générale. Néanmoins, le chef de projet doit dès le départ considérer que son activité a une vocation nationale donc refuser tout particularisme régional. Il est aidé d'une équipe d'analystes-programmeurs et entouré d'un "groupe de conception" chargé de rédiger le document de "définition des objectifs globaux" puis le "cahier des charges" de l'application, qui sont adressés pour avis à tous les services utilisateurs potentiels et aux chefs de projet des autres applications.
Le groupe de conception comprend 6 à 10 personnes représentant les services les plus divers concernés par le projet,et comporte obligatoirement un bon analyste attaché à l'application.

II - L'IMPLANTATION GEOGRAPHIQUE D'UN RESEAU INFORMATIQUE PERFORMANT

L'organisation de l'entreprise française des télécommunications repose sur l'existence de 20 régions. Des calculateurs ont été implantés dans le passé au moins dans toutes les plus importantes. On trouve ainsi des machines Bull Gamma 30 à Lyon et Marseille, des GE 425 à Lille, Bordeaux, Toulouse et Montpellier, un GE 437 à Massy, enfin quelques machines Bull 300 TI à programmes câblés étaient récemment ou sont encore en service dans les régions de Nancy, Nantes, Limoges, Poitiers et Rouen ; ce parc est essentiellement utilisé pour la comptabilité téléphonique.
A l'avenir, si la plupart des fichiers nécessaires aux applications décrites plus haut peuvent être gérés en temps différé, un certain nombre d'entre eux devront nécessairement être accessibles, voire mis à jour en temps réel : parmi ces derniers le fichier commercial des abonnés, le fichier des renseignements, le fichier des circuits, le fichier technique des abonnés contiendront des quantités considérables d'informations.
Le volume total de caractères à gérer en phase finale sur un ordinateur ayant en charge quelque 500 000 abonnés a été estimé à un milliard de caractères au moins. Au moins le tiers des données seront concernées par des traitements en temps réel.
Aucun des calculateurs énumérés plus haut ne permettait d'envisager de tels traitements.
L'intégration progressive de toutes les applications suppose la création d'un support commun pour toutes les informations, une véritable "Banque de données", répartie sur des moyens de traitement nationaux et régionaux, et qui devra rester alimentée, mise à jour en permanence, à partir de la base de l'entreprise, c'est-à-dire les chantiers, les magasins, les guichets des services d'abonnement, les services de personnel etc.
L'étude des différents fichiers à constituer a donc permis de définir les principales caractéristiques du réseau d'ordinateurs nouveaux à mettre en place pour aborder la réalisation du système informatif. L'obligation de faire appel à des ordinateurs de troisième generation, très puissants et dotés de volumineuses mémoires de masse, a conduit à en réduire substantiellement le nombre.
L'implantation de sept centres de calcul interrégionaux constituera un compromis entre : d'une part le désir de réduire le coût économique de l'ensemble, de faciliter la coordination des équipes d'informaticiens; et d'autre part le refus de créer des centres trop importants difficiles à gérer et à diriger, et posant des problèmes délicats de sécurité. Le regroupement des traitements relatifs à plusieurs régions sur chacun de ces sept centres permettra de leur donner une taille relativement homogène. Chaque centre "gérera" environ un million d'abonnés à la fin du VIème Plan.
La mise en place de ces centres a débuté au début de l'année 1971 : un ordinateur IRIS 50 de la Compagnie Internationale pour l'Informatique a été installé à Toulouse en février ; la même machine vient d'être mise en service au centre de calcul interrégional de Bordeaux.

Figure 2.8.1.5 Print previous line (7.7 lines/mm).

L'ordre de lancement et de réalisation des applications fait l'objet de décisions au plus haut niveau de la Direction Générale des Télécommunications. Il n'est certes pas question de construire ce système intégré "en bloc" mais bien au contraire de procéder par étapes, par paliers successifs. Certaines applications, dont la rentabilité ne pourra être assurée, ne seront pas entreprises. Actuellement, sur trente applications qui ont pu être globalement définies, six en sont au stade de l'exploitation, six autres se sont vu donner la priorité pour leur réalisation.
Chaque application est confiée à un "chef de projet", responsable successivement de sa conception, de son analyse-programmation et de sa mise en oeuvre dans une région-pilote. La généralisation ultérieure de l'application réalisée dans cette région-pilote dépend des résultats obtenus et fait l'objet d'une décision de la Direction Générale. Néanmoins, le chef de projet doit dès le départ considérer que son activité a une vocation nationale donc refuser tout particularisme régional. Il est aidé d'une équipe d'analystes-programmeurs et entouré d'un "groupe de conception" chargé de rédiger le document de "définition des objectifs globaux" puis le "cahier des charges" de l'application, qui sont adressés pour avis à tous les services utilisateurs potentiels et aux chefs de projet des autres applications. Le groupe de conception comprend 6 à 10 personnes représentant les services les plus divers concernés par le projet,et comporte obligatoirement un bon analyste attaché à l'application.

II - L'IMPLANTATION GEOGRAPHIQUE D'UN RESEAU INFORMATIQUE PERFORMANT

L'organisation de l'entreprise française des télécommunications repose sur l'existence de 20 régions. Des calculateurs ont été implantés dans le passé au moins dans toutes les plus importantes. On trouve ainsi des machines Bull Gamma 30 à Lyon et Marseille, des GE 425 à Lille, Bordeaux, Toulouse et Montpellier, un GE 437 à Massy, enfin quelques machines Bull 300 TI à programmes câblés étaient récemment ou sont encore en service dans les régions de Nancy, Nantes, Limoges, Poitiers et Rouen ; ce parc est essentiellement utilisé pour la comptabilité téléphonique.
A l'avenir, si la plupart des fichiers nécessaires aux applications décrites plus haut peuvent être gérés en temps différé, un certain nombre d'entre eux devront nécessairement être accessibles, voire mis à jour en temps réel : parmi ces derniers le fichier commercial des abonnés, le fichier des renseignements, le fichier des circuits, le fichier technique des abonnés contiendront des quantités considérables d'informations.
Le volume total de caractères à gérer en phase finale sur un ordinateur ayant en charge quelques 500 000 abonnés a été estimé à un milliard de caractères au moins. Au moins le tiers des données seront concernées par des traitements en temps réel.
Aucun des calculateurs énumérés plus haut ne permettait d'envisager de tels traitements.
L'intégration progressive de toutes les applications suppose la création d'un support commun pour toutes les informations, une véritable "Banque de données", répartie sur des moyens de traitement nationaux et régionaux, et qui devra rester alimentée, mise à jour en permanence, à partir de la base de l'entreprise, c'est-à-dire les chantiers, les magasins, les guichets des services d'abonnement, les services de personnel etc.
L'étude des différents fichiers à constituer a donc permis de définir les principales caractéristiques du réseau d'ordinateurs nouveaux à mettre en place pour aborder la réalisation du système informatif. L'obligation de faire appel à des ordinateurs de troisième génération, très puissants et dotés de volumineuses mémoires de masse, a conduit à en réduire substantiellement le nombre.
L'implantation de sept centres de calcul interrégionaux constituera un compromis entre : d'une part le désir de réduire le coût économique de l'ensemble, de faciliter la coordination des équipes d'informaticiens; et d'autre part le refus de créer des centres trop importants difficiles à gérer et à diriger, et posant des problèmes délicats de sécurité. Le regroupement des traitements relatifs à plusieurs régions sur chacun de ces sept centres permettra de leur donner une taille relativement homogène. Chaque centre "gérera" environ un million d'abonnés à la fin du VIème Plan.
La mise en place de ces centres a débuté au début de l'année 197[illegible] : un ordinateur IRIS 50 de la Compagnie Internationale pour l'Informatique a été installé à Toulouse en février ; la même machine vient d'être mise en service au centre de calcul interrégional de Bordeaux.

Figure 2.8.1.6 Previous line/white (7.7 lines/mm).

L'ordre de lancement et de réalisation des applications fait l'objet de décisions au plus haut niveau de la Direction Générale des Télécommunications. Il n'est certes pas question de construire ce système intégré "en bloc" mais bien au contraire de procéder par étapes, par paliers successifs. Certaines applications, dont la rentabilité ne pourra être assurée, ne seront pas entreprises. Actuellement, sur trente applications qui ont pu être globalement définies, six en sont au stade de l'exploitation, six autres se sont vu donner la priorité pour leur réalisation.
Chaque application est confiée à un "chef de projet", responsable successivement de sa conception, de son analyse-programmation et de sa mise en oeuvre dans une région-pilote. La généralisation ultérieure de l'application réalisée dans cette région-pilote dépend des résultats obtenus et fait l'objet d'une décision de la Direction Générale. Néanmoins, le chef de projet doit dès le départ considérer que son activité a une vocation nationale donc refuser tout particularisme régional. Il est aidé d'une équipe d'analystes-programmeurs et entouré d'un "groupe de conception" chargé de rédiger le document de "définition des objectifs globaux" puis le "cahier des charges" de l'application, qui sont adressés pour avis à tous les services utilisateurs potentiels et aux chefs de projet des autres applications. Le groupe de conception comprend 6 à 10 personnes représentant les services les plus divers concernés par le projet,et comporte obligatoirement un bon analyste attaché à l'application.

II - L'IMPLANTATION GEOGRAPHIQUE D'UN RESEAU INFORMATIQUE PERFORMANT

L'organisation de l'entreprise française des télécommunications repose sur l'existence de 20 régions. Des calculateurs ont été implantés dans le passé au moins dans toutes les plus importantes. On trouve ainsi des machines Bull Gamma 30 à Lyon et Marseille, des GE 425 à Lille, Bordeaux, Toulouse et Montpellier, un GE 437 à Massy, enfin quelques machines Bull 300 TI à programmes câblés étaient récemment ou sont encore en service dans les régions de Nancy, Nantes, Limoges, Poitiers et Rouen ; ce parc est essentiellement utilisé pour la comptabilité téléphonique.
A l'avenir, si la plupart des fichiers nécessaires aux applications décrites plus haut peuvent être gérés en temps différé, un certain nombre d'entre eux devront nécessairement être accessibles, voire mis à jour en temps réel : parmi ces derniers le fichier commercial des abonnés, le fichier des renseignements, le fichier des circuits, le fichier technique des abonnés contiendront des quantités considérables d'informations.
Le volume total de caractères à gérer en phase finale sur un ordinateur ayant en charge quelques 500 000 abonnés a été estimé à un milliard de caractères au moins. Au moins le tiers des données seront concernées par des traitements en temps réel.
Aucun des calculateurs énumérés plus haut ne permettait d'envisager de tels traitements.
L'intégration progressive de toutes les applications suppose la création d'un support commun pour toutes les informations, une véritable "Banque de données", répartie sur des moyens de traitement nationaux et régionaux, et qui devra rester alimentée, mise à jour en permanence, à partir de la base de l'entreprise, c'est-à-dire les chantiers, les magasins, les guichets des services d'abonnement, les services de personnel etc.
L'étude des différents fichiers à constituer a donc permis de définir les principales caractéristiques du réseau d'ordinateurs nouveaux à mettre en place pour aborder la réalisation du système informatif. L'obligation de faire appel à des ordinateurs de troisième génération, très puissants et dotés de volumineuses mémoires de masse, a conduit à en réduire substantiellement le nombre.
L'implantation de sept centres de calcul interrégionaux constituera un compromis entre : d'une part le désir de réduire le coût économique de l'ensemble, de faciliter la coordination des équipes d'informaticiens; et d'autre part le refus de créer des centres trop importants difficiles à gérer et à diriger,et posant des problèmes délicats de sécurité. Le regroupement des traitements relatifs à plusieurs régions sur chacun de ces sept centres permettra de leur donner une taille relativement homogène. Chaque centre "gérera" environ un million d'abonnés à la fin du VIème Plan.
La mise en place de ces centres a débuté au début de l'année 197 : un ordinateur IRIS 50 de la Compagnie Internationale pour l'Informatique a été installé à Toulouse en février , la même machine vient d'être mise en service au centre de calcul interrégional de Bordeaux.

Figure 2.8.1.7 Normal decode/previous line (7.7 lines/mm).

for good quality transmission lines. The transmission time may also be reduced by error control because it may be possible to transmit at a higher bit rate. Error control also has the obvious advantage of eliminating the error streaks in the output copy.

The CCITT has recently completed its investigation of error control for Group 3, and the work is summarized in two different options to appear in the CCITT Blue Book published in 1989. One option merely limits the effect of errors, but does not correct them. The second option employs true error correction techniques.

The error limiting option applies only to one-dimensional MH coding. The first step is to divide the 1728 pels per line into 12 groups of 144 pels each. Each group is next classified to be either all white or nonwhite. A 12-bit code word at the beginning of each line identifies the white and nonwhite parts. The all-white groups are not encoded. The nonwhite groups are encoded separately by the standard MH. The intent of this option is to limit the effect of an error to half of a scan line rather than the entire scan line. The value of this option has yet to be proved in practice.

The error correction option is designed to be more universal in its application than the error limiting option, and has been implemented in many fax units. The design objectives for the error control option were that the number of changes in the existing CCITT Recommendations T.4 and T.30 should be minimized, it should operate well in a burst error environment, and the overhead for error-free channels should be low. This option applies to both the MH and MR modes of operation. The error control technique chosen by the CCITT is a half-duplex HDLC selectively repeating ARQ concept. The error control procedure is summarized below.

- The coded data stream is divided into HDLC "frames" of 256 bytes each.
- A redundant code is synthesized and transmitted within each frame which permits the receiver to detect a transmission error in that frame.
- After the page is completely transmitted, the receiver requests the retransmission of the frames in error.
- The transmitter sends the bad frames.
- After four requests for retransmission of the same frame, the transmitter may stop or continue with optional modem speed fallback.

2.8.3 Dither Coding

Over eight million facsimile units are now operational, and sales are continuing at an increasing rate. This success can be traced to many causes, two which are listed below.

- The scanned image is first converted to binary pixels (black or white), which are easily compressed to minimize transmission time.

- The printer is a binary device that writes only black or white dots. This has permitted the printer to be produced with high reliability and very low cost.

Most of the images transmitted through fax systems contain only black or white information (text, line drawings, *et cetera*). Because the transmission system and printing system are binary, the output quality for documents of this type is excellent. When a gray scale image is transmitted, however, the output image is usually severely distorted. The purpose of dither coding is to transmit gray scale images through facsimile systems employing binary transmission and binary printing techniques with improved output quality.

The Basic Dither Process

The threshold circuit in a bi-level facsimile system is the key element in dither coding. For purposes of analysis, it can be assumed that conventional bi-level systems employ a fixed threshold at the midpoint between peak black and peak white. Figure 2.8.3.1 (a) illustrates such a coding process. Note that an input gray level near the threshold is drastically altered in the output image. In dither coding, the threshold is varied, or dithered, in amplitude from pel to pel, as shown in Figure 2.8.3.1 (b). If the threshold is dithered uniformly over the gray scale range, the average value of the output image over a number of neighboring pixels will approximate the input gray value; the eye will tend to perform the averaging function, and the observer will perceive the input gray scale. A computer simulation study of dither coding has been performed in which clumped dither, ordered dither, and two enhancements were investigated. The results of this study are provided below. (See Table 2.8.3.1)

Clumped Dither

The clumped dither compression technique is an electronic approximation of the photo-mechanical screening process. It employs a matrix of fixed thresholds, shown in Figure 2.8.3.2, which are arranged in such a way that a "dot" grows outward from the center as successively darker gray levels are encountered in low contrast regions of the image.* The numbers in the matrix represent threshold values out of 256 possible brightness levels. The eight-bit gray level of each input pel is compared to one of the matrix thresholds; which threshold is used depends upon the position of the pel in the image. Figure 2.8.3.3 shows how the irregular matrices interlock.

* J. C. Stoffel, *Graphical and Binary Image Processing and Applications,* Artech House, Norwood, MA, 1982, p. 295.

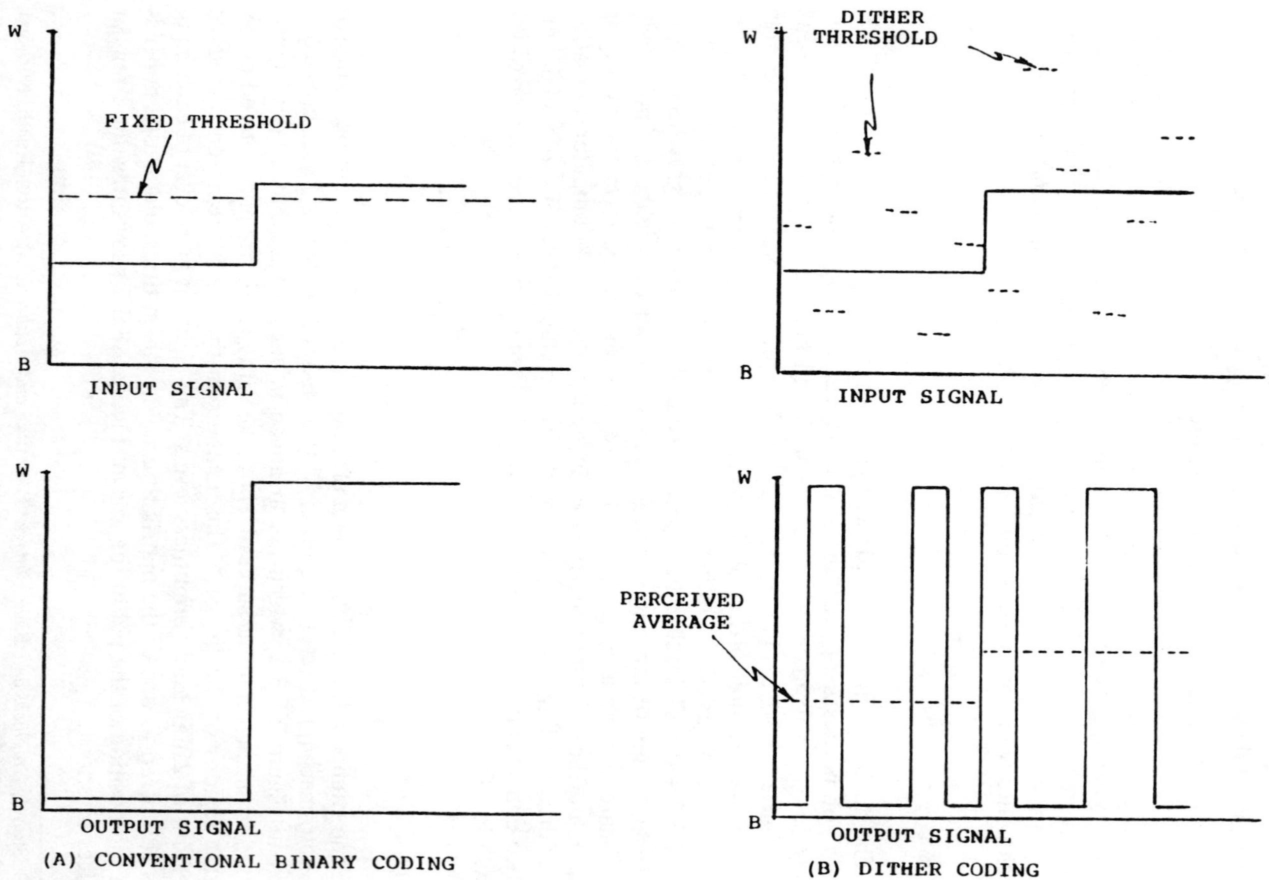

Figure 2.8.3.1 Dither coding: **(a)** conventional binary coding; **(b)** dither coding.

Table 2.8.3.1. Compression Results for Dither Coding Techniques
(Bits/Pixel)

Coding Technique	*House with Sky*	*House with Trees*	*Aerial Photo*
Clumped	0.98	0.81	0.98
Clumped with plane coding	0.40	0.66	0.77
Ordered	1.25	0.99	1.10
Ordered with reordering	0.47	0.52	0.52

Clumped Dither with Plane Coding

In the clumped dither with plane coding technique, the dithered image is separated into 17 planes, where each plane is comprised of the pixels associated with one of the thresholds in the clumped dither matrix. The pixel in matrix position 1 of every clumped dither matrix is placed in plane 1, the pixel in matrix position 2 of every clumped dither matrix is placed in plane 2, and so on. Placing together all of the pixels compared to the same threshold, in most cases, increases the length of black or white runs, thus making the image more compressible. The 17 clump planes are then compressed by using Group 4 encoding. At the receiver, the planes are all decoded, and the dithered image is reconstructed from the clump planes by placing each pixel back in the position from which it was extracted at the transmitter. Clump plane separation provides enhanced compression, in most cases, with no effect on image quality. This technique, however, is more complex to implement than conventional clumped dither.

Ordered Dither

Ordered dithering employs an $n \times n$ matrix of fixed thresholds that is repeated throughout the image, in the same way as clumped dithering. The distribution of the thresholds within the matrix is designed to provide acceptable pseudogray scale rendition and edge sharpness while producing a minimum of visible patterns or artifacts. The ordered dither matrix employed in this study is an 8×8 matrix in which the thresholds are arranged symmetrically, as shown in

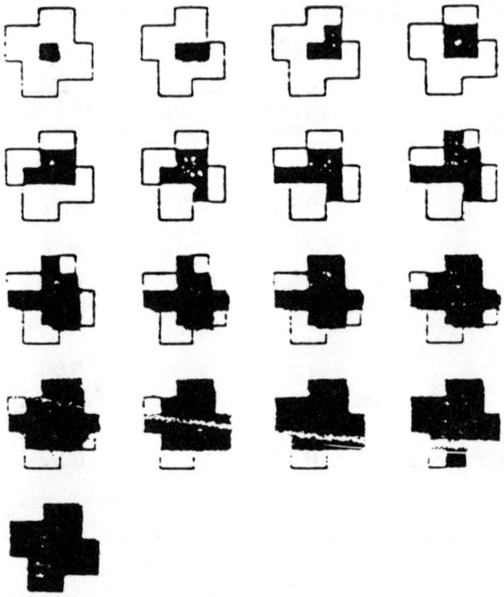

(a) Halftone dots in nondetail regions.

		167	200	
230	210	94	72	
153	111	36	52	193
	216	181	126	222
	242	232		

(b) Clumped dither matrix.

Figure 2.8.3.2

Figure 2.8.3.4. The threshold values are related to a maximum of 32 shades of gray from black to white. The even-numbered thresholds (2–30) are arranged spirally in the upper right and lower left corners, and the odd-numbered thresholds

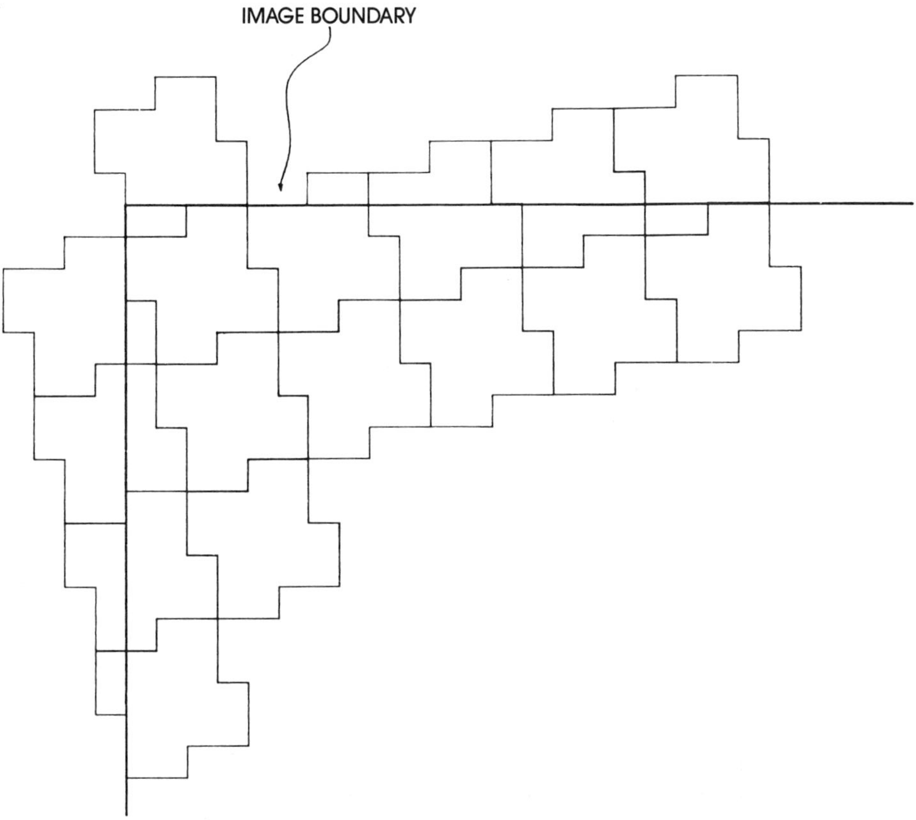

Figure 2.8.3.3 Clumped dither matrix lattice.

(1–31) are arranged spirally in the upper left and lower right corners. This arrangement provides 32 pseudogray scale levels with a pattern of four "dots" growing in each of the four corners of the matrix as successively darker gray levels are encountered in low contrast regions of the image.

Ordered Dithering with Bit Reordering

If the ordered dither pixels, described above, are transmitted in the conventional raster pattern as shown in Figure 2.8.3.4, the length of the black and white rows usually will be short, and the compression will be poor. The purpose of the reordering technique is to transmit the pixels in a sequence where the contiguous transmitted pixels are more correlated, and therefore more compressible. Figure 2.8.3.5 illustrates the reordering process.

25	19	11	27	24	18	10	2
9	1	3	21	8	1	2	20
7	7	5	13	16	6	4	12
31	15	23	29	30	14	22	28
24	18	10	26	25	19	11	27
8	1	2	20	9	1	3	21
16	6	4	12	17	7	5	13
30	14	22	18	31	15	23	29

Figure 2.8.3.4 Original matrix.

1	3	5	7	6	4	2	1
9	11	13	15	14	12	10	8
17	19	21	23	22	20	18	16
25	27	29	31	30	28	26	24
24	26	28	30	31	29	27	25
16	18	20	22	23	21	19	17
8	10	12	14	15	13	11	9
1	2	4	6	7	5	3	1

Figure 2.8.3.5 Reordered matrix.

Test Results

The compression performance of the four dither coding techniques described above was measured by means of computer simulation. The compression algorithm used in the simulation is the modified modified read (MMR) coding technique used in Group 4 facsimile equipment. The three images shown in Figure 2.8.3.6 were used in the simulations, and the compression results are summarized in Table 2.8.3.1. Output images were generated for the two coding techniques and the quality was found to be quite similar.

Figure 2.8.3.6 Test images.

2.8.4 RS-232 Interface

Figure 2.8.4.1 is a functional block diagram of the typical facsimile machine indicating that the only I/O connection is a modem to interface with the PSTN. Table 2.8.4.1 indicates the three possible modes of operation for a conventional

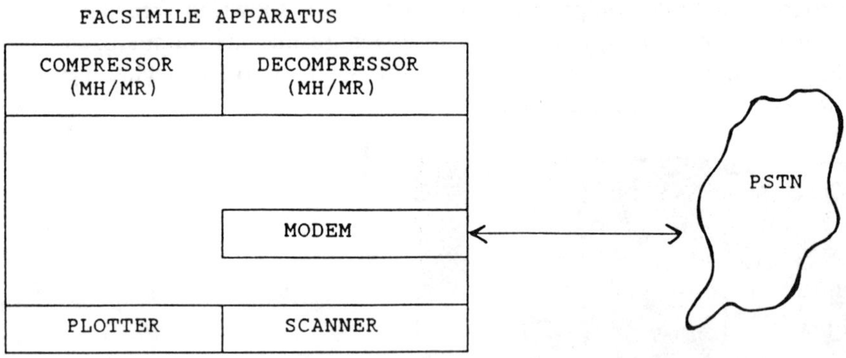

Figure 2.8.4.1 Typical facsimile machine.

Table 2.8.4.1. Functions of Facsimile Machines with RS-232-C

	Function	*Source*	*Destination*	*Input*	*Output*
Without	Fax Transmit	Scanner	Modem	Image	MH/MR
RS-232-C	Fax Receive	Modem	Plotter	MH/MR	
	Copy	Scanner	Plotter	Image	Image
	Plot	RS-232-C	Plotter	MH/MR/Image	Image
	Scan	Scanner	RS-232-C	Image	
RS-232-C	Transmit	RS-232-C	Modem	MH/MR/Image	MH/MR
	Receive	Modem	RS-232-C	MH/MR	
	Compress	RS-232-C	RS-232-C	Image	MH/MR
	Decompress	RS-232-C	RS-232-C	MH/MR	
RS-232-C	Print	RS-232-C	Plotter	ASCII	Image
plus Character	Remote Print	RS-232-C	Modem	ASCII	MH/MR
Generator	Create Image	RS-232-C	RS-232-C	ASCII	MH/MR/Image

facsimile system: transmit, receive, and copy. Many facsimile manufacturers are offering an optional digital interface to connect directly with a local computer, digital network, cryptographic device, local scanner, or external modem. This digital interface is invariably accomplished in accordance with the RS-232 standard. Figure 2.8.4.2 is a functional block diagram illustrating the digital configuration with the RS-232 interface. Table 2.8.4.1 lists six different modes of operation that can be accomplished by using the digital interface. In the "transmit" mode, imagery that originates in a computer is fed to the fax unit for transmission. In the "receive" mode, pictures received by the fax unit are fed directly to a computer, LAN, or cryptographic machine.

Another typical option frequently associated with the RS-232 mode is the inclusion of a character generator as shown in Figure 2.8.4.2 and Table 2.8.4.1. In the "remote print" mode, the fax unit accepts ASCII character data from the local digital source, the character generator converts the data to raster format, and the MH/MR compressor prepares this signal for conventional fax transmission over the PSTN. The advantage of this mode as well as the "transmit" mode is that the input image is never scanned, and therefore the output image does not have the degradation associated with the scanning process.

It is unfortunate that no standard exists to define the signal protocol and frame which pass over the RS-232 interface. Consequently, fax units from different manufacturers are not compatible over the digital interface.

2.8.5 High Resolution

The nominal resolution for Group 3 facsimile in the horizontal direction is 200 pels/inch. A large fraction of Group 3 fax transmissions employ the high resolution option, which is 200 scan lines/inch. Therefore, in this mode, the pels are essentially square: 200 pels/inch in both directions. There have been discussions about the possibility of extending the resolution of Group 3 fax beyond this present capability of 200 pels/inch. As part of this evaluation it is useful to consider how the compression factor is affected by an increase in resolution.

Figure 2.8.5.1 is a plot of the number of bits/page at 200 pels/inch, and higher resolutions, for three of the standard CCITT documents. Notice that in each case the number of bits/page at 400 pels/inch is approximately double that at 200 pels/inch. This occurs in spite of the fact that there are four times as many pels in the 400 pel/inch image relative to the 200 pel/inch image. This means that the compression ratio increases as the resolution increases, as shown in Figure 2.8.5.2. In general the compression ratio doubles as the resolution doubles. The compression increases as the resolution increases due to the higher level of correlation and redundancy between adjacent pels.

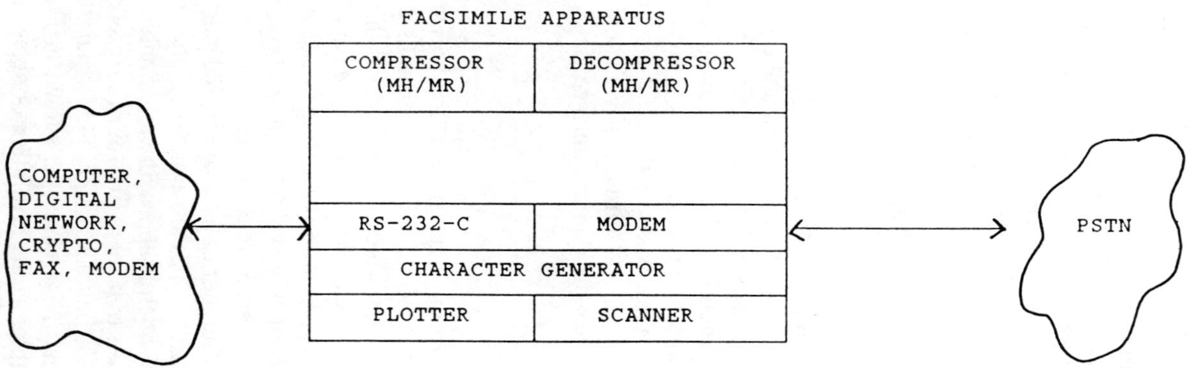

Figure 2.8.4.2 Facsimile machine with RS-232-C.

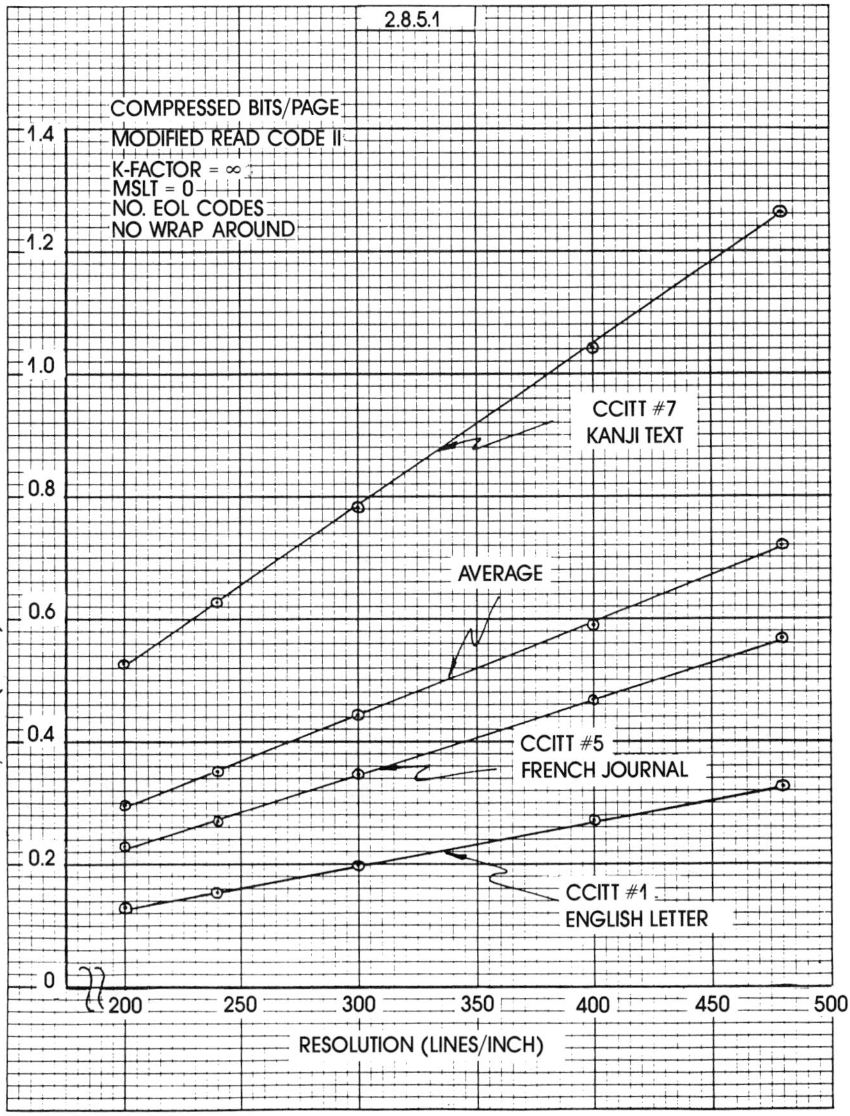

Figure 2.8.5.1 Bits per page (× 10^6).

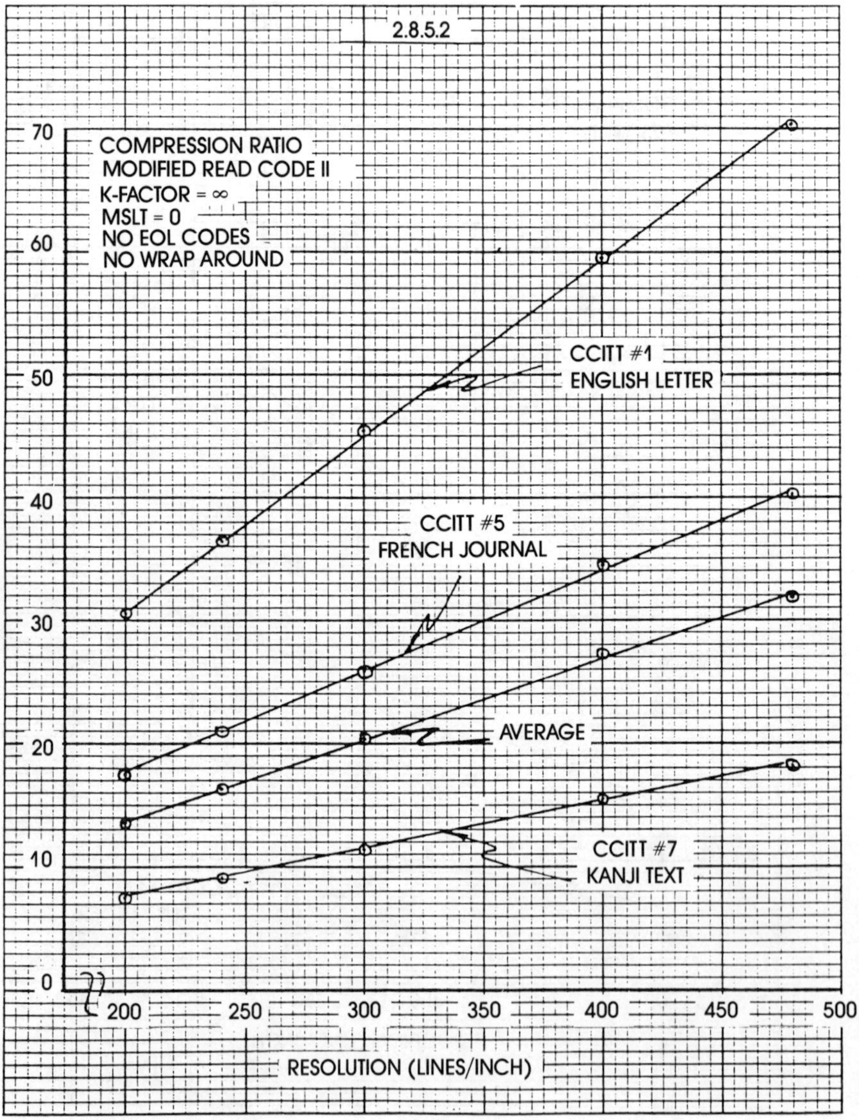

Figure 2.8.5.2 Compression ratio.

2.8.6 Small Page Size

The basic Group 3 standard is defined for the scanning and printing of an A4 page which is 215 mm (8.5 inches) wide. A new CCITT recommendation in the 1989 Blue Book defines two other optional page sizes that may be used for Group 3 operation. An A5 page is 151 mm (5.9 inches) wide and the A6 page is 107 mm (4.2 inches) wide.

The primary purpose of these reduced page sizes for fax is to permit the development of small, inexpensive devices such as those which can operate with a telephone.

The CCITT recommendation not only defines the size of the input-output pages, but also the resolution which is to be used for various page sizes. Figure 2.8.6.1 is from the CCITT recommendation defining the resolution in numerous modes of operation. The reader will note that it is permissible to mix page sizes, to transmit from any input page size (e.g., A4) to any other page size (e.g., A6). When different page sizes are used, there will be a corresponding reduction or enlargement at the output to ensure that no data are lost. Figure 2.8.6.1 defines system operation for all combinations of input and output page sizes. See Table 2.7.6.1 for the DIS and DTC bit settings to allow these new small page units to operate in one or more possible modes.

Provision is made for sending from a standard width (A4) transmitter with 8 pels/mm across an 8.5-inch wide page to an A6 receiver with 16 pels/mm across a 4.25-inch wide page. All information would be preserved, but the user might need a magnifying glass to read the small print size. Cost of the electronics required would probably be the same as for the A4 fax unit. An A5 fax unit that preserved all of the information would record on 7-inch wide paper. No changes would be needed in existing Group 3 fax units to send in this manner to A5 or A6 fax units.

Alternatively, sending from an A4 transmitter to an A5 or A6 receiver can be done by throwing away pels at the transmitter to match the number of pels that the receiver prints. This method is similar to sending from A3 to A4 fax units. Figure 2.8.6.2. shows that copy received by this method has lower resolution.

Existing Group 3 fax units would not be able to throw away pels to match A5 or A6 requirements. The transmitter, however, could send a full line of 1728 pels and the receiver could throw away the extra pels.

Sending from A5 to A5 or A6 to A6 fax units keeps the same narrower copy size without enlargement or reduction, providing both fax units have the same number of pels per line. (See Figure 2.8.6.3.)

Sending from A5 or A6 fax units to an A4 unit can be done with the narrow page text centered on the A4 page.

Interworking between equipments with A5/A6 and A4 facilities and between equipments with combinations of these facilities

Situation at transmission side / Terminal capabilities at reception side				§ 2.1 case	b)	e)	g)	f)	h)
				Horizontal resolution	1728 pels/ 215 mm	864 pels/ 107 mm	1728 pels/ 107 mm	1216 pels/ 151 mm	1728 pels/ 151 mm
				Vertical resolution	3.85 l/mm 7.7 l/mm	7.7 l/mm 15.4 l/mm	7.7 l/mm 15.4 l/mm	5.44 l/mm 10.9 l/mm	5.44 l/mm 10.9 l/mm
				Pel process	Original 1728	864 1728 * .5) Note 1	Original 1728	1216 1728 * .70 Note 2	Original 1728
§ 2.1 case	Horizontal resolution	Vertical resolution	Pel process	DIS-DTC: / DCS	–	Bit 33 = 1 Bit 35 = 1	Bit 33 = 1 Bit 37 = 1	Bit 33 = 1 Bit 34 = 1	Bit 33 = 1 Bit 36 = 1
b)	1728 pels/ 215 mm	3.85 l/mm 7.7 l/mm	Original 1728		Equal (A4)	Reduced (A4 → A6)	Reduced (A4 → A6)	Reduced (A4 → A5)	Reduced (A4 → A5)
e)	864 pels/ 107 mm	7.7 l/mm 15.4 l/mm	864 * 2 Note 1	Bit 17 = 0 Bit 18 = 0 Rit 33 = 0	Enlarged (A6 → A4)	Equal (A6) Note 1	Equal (A6)	Enlarged (A6 → A5)	Enlarged (A6 → A5)
g)	1728 pels/ 107 mm	7.7 l/mm 15.4 l/mm	Original 1728		Enlarged (A6 → A4)	Equal (A6)	Equal (A6)	Enlarged (A6 → A5)	Enlarged (A6 → A5)
f)	1216 pels/ 151 mm	5.44 l/mm 10.9 l/mm	1216 * 1.42 Note 2	Note 1, 2	Enlarged (A5 → A4)	Reduced (A5 → A6)	Reduced (A5 → A6)	Equal (A5) Note 2	Equal (A5)
h)	1728 pels/ 151 mm	5.44 l/mm 10.9 l/mm	Original 1728		Enlarged (A5 → A4)	Reduced (A5 → A6)	Reduced (A5 → A6)	Equal (A5)	Equal (A5)

Note 1 – Bit 33 = 1 Transmit pel process = 432(W) + 864 + 432 (W)

Bit 35 = 1 Receive pel process extracts central 864 pels

Note 2 – Bit 33 = 1 Transmit pel process = 256(W) + 1216 + 256(W)

Bit 34 = 1 Receive pel process extracts central 1216 pels

(W) = white pels

Figure 2.8.6.1 Recommended resolution.

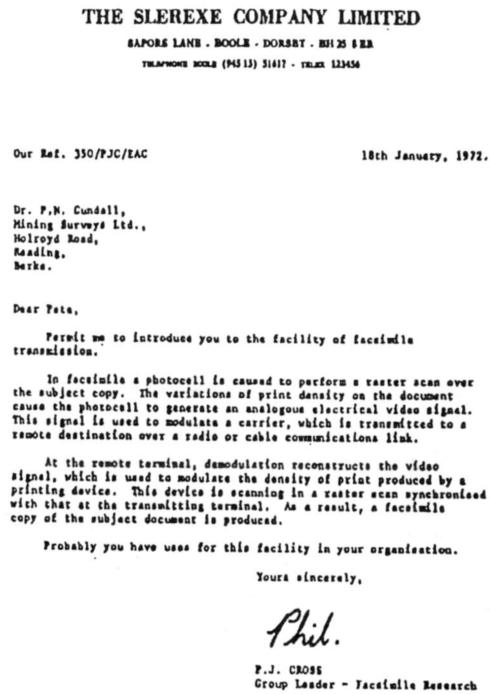

THE SLEREXE COMPANY LIMITED

SAPORS LANE . BOOLE . DORSET . BH 25 8 ER

TELEPHONE BOOLE (945 13) 51617 - TELEX 123456

Our Ref. 350/PJC/EAC 18th January, 1972.

Dr. P.N. Cundall,
Mining Surveys Ltd.,
Holroyd Road,
Reading,
Berks.

Dear Pete,

Permit me to introduce you to the facility of facsimile transmission.

In facsimile a photocell is caused to perform a raster scan over the subject copy. The variations of print density on the document cause the photocell to generate an analogous electrical video signal. This signal is used to modulate a carrier, which is transmitted to a remote destination over a radio or cable communications link.

At the remote terminal, demodulation reconstructs the video signal, which is used to modulate the density of print produced by a printing device. This device is scanning in a raster scan synchronised with that at the transmitting terminal. As a result, a facsimile copy of the subject document is produced.

Probably you have uses for this facility in your organisation.

Yours sincerely,

Phil.

P.J. CROSS
Group Leader - Facsimile Research

Figure 2.8.6.2 Received copy — A6 from A4.

As the facsimile service becomes increasingly oriented toward the consumer with very low cost terminal devices, this small page option may be employed more frequently.

2.8.7 Nonstandard Operation

The Group 3 facsimile standard is specifically designed to permit a fax transmitter to request that the receiver switch to a nonstandard mode of operation. If the receiver is equipped with the appropriate proprietary capability, transmission can occur in the nonstandard mode. This nonstandard mode of operation, which is permitted and defined by the Group 3 recommendation, permits fax manufacturers to provide improved technology and service to users beyond those which

SAMPLE DOCUMENT
(MT 5 TEST CHART NO.2)

IF you believe that their marketing
organization could be of help to you
in your current business, then this
could be an excellent deal where you
could not possibly lose and where there
would be a lot to gain. You could pay
th book value down in your stock with
the additional half a million contingent
upon future earnings of Industries.
John Alden is a wealthy individual and the
kind of a guy that would be interested in a
deal with our stock.
I would urge you to have your accountants
look over the financials immediately and
come to some intelligent understanding as
to just what the picture looks like.

Figure 2.8.6.3 Received copy — A6 from A6.

have been provided by the standard. For example, several manufacturers have provided a proprietary mode of operation to users which employs nonstandard compression algorithms and error control techniques.

2.8.8 Group 3 with Group 4–like Features

The protocols for Group 4 were developed for compatibility with teletex and other telematic services. Teletex was planned as a modern replacement for telex as the record communication terminal that would be dominant for business communication. The CCITT teletex recommendations were issued in 1980, the same year as Group 3 fax. Contrary to expectations, teletex is hardly used by the business community except in the Federal Republic of Germany. It appears now that there is no reason to have the highly complex protocols which were evolved for Group 4. Work has continued on CCITT Group 4 recommendations. They are complex and overspecified. High user costs, an increasingly complex protocol, and limited availability of ISDN-like channels continue to delay this program. Group 4 classes

2 and 3 are of no use without a large number of teletex units with which to communicate.

Proposals are being discussed for Group 3 enhancements to provide desirable Group 4 features such as high resolution, error-free transmission, and high speed without incurring its high costs. The features could be introduced under accelerated procedures, the way that A5-A6 and the error-free proposals were handled. The first simplification would be to eliminate requirements to interface with teletex. The Group 4 teletex connection greatly delayed Group 4 and unduly complicated the design and protocols.

A lower cost alternative for operation on the new integrated services digital network (ISDN) is possible by expanding T.30 for Group 3 operation on it. A digital rate of 64 kb/s could be used with one of the B channels, or an analog interface could be used in the same manner as voice is handled. Error correction could be based on the selective repeat option, now a part of T.30. Full-duplex operation could be used to speed up the error-correction process. Optional resolutions of 300 and 400 lines/inch would give the higher resolutions which many Group 4 users want. Alternative higher modem speeds could be utilized on the PSTN (14,400 or 19,200 b/s) to reduce sending time.

2.8.9 Fax-Phone Switch Boxes

Fax performs best when a separate telephone line is dedicated to its use. This way, someone talking on the phone will not delay reception of important fax messages. Users who have only a small amount of fax traffic, however, may want to consider a fax-phone switch box to avoid the need for another line. When a call is received, the switch box connects the line to either the fax or the phone. The simplest design merely counts rings and switches the line to the fax unit if it is not answered for a voice call before the preset number of rings. Many fax units can be programmed to answer after a preset number of rings, making the box unnecessary.

Other switch boxes listen for CNG, the 1100-Hz beep tone from an automatic calling fax unit, for 4 seconds or up to 10 seconds and switch the receiving fax unit to the line when this tone is heard. If the phone is picked up, a button on the box allows the line to be switched. For a manual calling fax unit, the person who answers the phone pushes the fax switching button or the switch box will switch the unit to the line automatically after a preset number of rings.

More sophisticated switch boxes have a digital voice announcement telling the caller which touch-tone phone button to push for fax or leaving a voice message. Others connect the customer's voice answering machine to give the same type of message. Some switch boxes have a 3- or 4-digit security code to screen unwanted calls. The timing is important to allow the desired function to happen before the

35-second timeout disconnect of a Group 3 fax unit. A fax voice switch box may cost $300.

2.9 COMPUTER-BASED FACSIMILE

2.9.1 PC-Fax Cards

A PC-Fax card, installed in a vacant slot in a personal computer (PC) with suitable software, allows it to emulate a Group 3 fax unit. The keyboard operator selects commands from a screen menu to go through the steps needed to transfer text or image files. Like a Group 3 fax, the PC automatically negotiates the highest sending speed possible, then sends. More time is needed before starting to send, however, than with a Group 3 fax.

A document is composed and edited on the computer screen in the normal manner to make a computer file. Improved TSR (terminate and stay resident) software allows a "hot key" to call up the PC-fax program directly from the computer applications program in use. Menu-driven screens guide the user through the necessary steps. The telephone number of the receiving fax unit is selected from a directory stored in the computer. Word processing, ASCII, and other computer files can be converted to Group 3 format while being sent as a fax signal to the selected Group 3 fax unit. When sending fax from a PC, the image sent does not exist as hard copy. (See Figure 2.9.1.1.)

The items in the following description of PC-fax cards are not always furnished by some PC-fax card designs. The modem incorporated on many PC-fax cards is the same CCITT V.29, V.27(ter), V.21 modem used in Group 3 fax units. It is programmed to attempt first to send at 9600 b/s, with the ability to step down to 7200, 4800, or 2400 b/s, if required by the telephone line characteristics.

When a PC-fax receives from another PC-fax or a Group 3 fax unit, the document is stored as a file on a computer disk. Received files may be viewed from the computer screen display as soft copy, or a hard copy can be made on a dot matrix, laser, or other bit-mapped printer.

PC-fax units compress the image with MH coding before storage. A page in storage typically takes 50 to 60 K of memory, but up to 100 K may be required. While a hard disk is not necessary for this service, only a few pages can be stored without one. The computer can organize the documents in memory by dates, subjects, or originator to simplify search and retrieval. Background operation of the computer allows PC-fax to receive facsimile copies without interruption of other work being done by the computer.

One or more directory lists of telephone numbers can be stored in the computer for fax transmission. These numbers can easily be retrieved for automatic calling for immediate or delayed transmission of fax documents. Delayed trans-

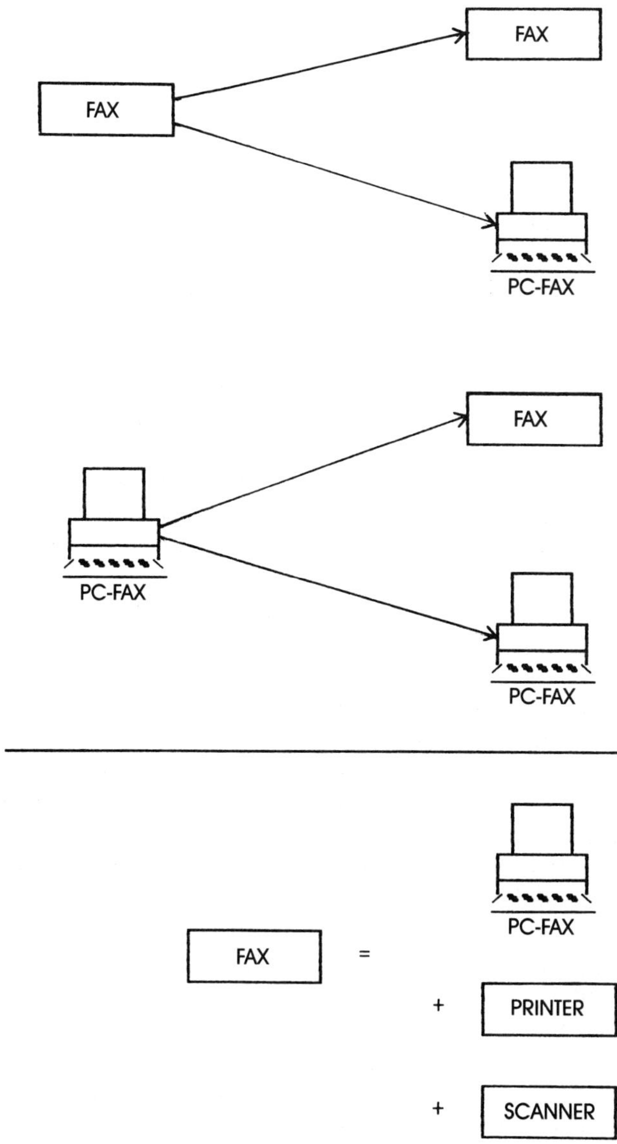

Figure 2.9.1.1 Use of PC-fax.

mission may be specified to send at the time of lowest telephone line charges. Sequential calling of a group of fax units may be used for automatically sending identical documents.

Security against unwanted viewing of the received document is good on a PC-fax. On most Group 3 fax units, the pages are stacked in a tray for any passerby to pick up and read. Some of the more expensive fax units have a mailbox option, where the document is kept in memory until accessed by password.

To prevent unauthorized interception and printing, the fax signal can be encrypted as it is passed from the applications program to disk storage. An intercepted signal could not be decoded without the proper encryption key. Finding the proper key by trial and error is almost impossible. Encryption could be incorporated when sending between PC-fax units, but it might be difficult to match encryption systems between a PC-fax unit and a Group 3 fax unit.

Although some of the less expensive PC-fax units require conversion of files to Group 3 format before dialing the fax receiver, many PC-fax units will send ASCII files directly as Group 3 fax files. Non-ASCII word-processing, spreadsheet, and graphic files may be sent in a similar manner. From the menu selection, the PC is given the "print" command for a particular printer specified by the PC-fax program. The fax signal addressed to the printer is redirected to a disk file. It has been stripped of control characters and stored as an ASCII file, which is then sent in Group 3 fax format.

To have a complete Group 3 fax replacement, the PC-fax must have a scanner to read information from hard-copy originals. Scanners generally have 300 dots per inch (dpi) to match the resolution of the popular laser printers. If the scanner resolution is higher than the 203 dpi of Group 3 fax, the image file can be converted by the PC to this resolution. With a scanner, the user can enter headers and signatures into computer memory. These files may be chained to a text file to generate letters, reports, and other business documents which look like they were typed on letterhead stationery and signed before sending. Programs such as PC Paintbrush™ or Dr. Halo™ can be used to edit fax image files displayed on the PC screen. These programs can be used to compose a page combining graphics and text before sending. They also can be used to modify received fax documents.

The price of a page scanner is often more than an inexpensive Group 3 fax unit. Purchase of a supplementary fax unit may be a better investment than a scanner for some PC-fax board users. The fax unit can be used as a scanner for the PC and can also print out edited image files. Sharpness of the recorded copy is better on the Group 3 fax unit than on the best quality 7 × 9 dot matrix printer.

The received copy quality is better when sent from a PC-fax card than a Group 3 fax unit. The PC-fax uses a character generator to place each dot in the matrix for a character in the optimum position for the typeface used. When a fax unit scans a character, the dot matrix formed depends on the exact character position related to the scan. The black dot pattern sent for a given character scanned a number of times will have different nonoptimum renditions. Figure 6.2.4 is a computer printout of a Group 3 scanned image. Notice that none of the fax printed letters "a" have the same pattern.

Integrated circuit (IC) chips for compressing the facsimile signals have both Group 3 and 4 standards, giving the capability for resolutions of 200, 300, and 400 dpi. Some fax boards incorporate 300 dpi as an alternative Group 3 system resolution. The 300-dpi laser and ink-jet printers or high-resolution dot matrix printers provide a better resolution copy between PC-fax board users when receiving documents scanned at 300 dpi. This feature, available now, could have a negative effect on the demand for Group 4 facsimile units.

A portable computer capable of mobile operation can be fitted with a PC-fax card for Group 3 communication in an automobile or other vehicle using a cellular phone system. See Section 9.2.5 for problems encountered when operating on cellular phone systems.

Personal computer users wishing to send text files have been hampered by the lack of a universally accepted PC data communication standard. Users now can easily emulate Group 3 fax to send information to the more than eight million (in 1988) Group 3 fax machines connected to the regular telephone network (PSTN). Most of these fax units are available 24 hours a day, like a regular telephone.

Some PC-fax systems allow binary file transfer of computer data between PCs at 9600 b/s with the high-speed fax modems. This capability greatly reduces the telephone line charges compared to 1200 or 300 b/s with the commonly used Hayes-compatible modems. Incompatible protocols have been developed by each vendor for this purpose. An error-correction mode of sending computer data is very important because even a few errors cannot be tolerated. Work is underway for adapting the Group 3 fax error-correction mode for block transmission of computer data. The intent is to modify the CCITT Group 3 recommendation to add an option for compatibility between all PC-fax card users.

2.9.2 PC-Fax Controllers

The PC-fax controller is designed to broaden the use of Group 3 fax for the delivery of electronic mail (E-Mail) and telex messages. The controller also gives local area network (LAN) users and computer terminals the ability to access the eight million Group 3 fax units worldwide for delivery of ASCII and graphics data. PC-fax controllers, however, are not designed to substitute for a PC-fax card. Multiple users can access the controller through a gateway server to a LAN network, a mainframe computer, or a concentrator for personal computers. As an example, a London bank has 25 LAN networks bridged to provide Group 3 fax sending capability for 200 terminals through a PSTN communication server with multiple telephone lines.

A typical PC-fax controller might access a mainframe computer through an RS-232 port. Each terminal connected to the mainframe has the ability to send to Group 3 fax units through the controller. It has all of the functions of a high-level PC-fax card plus others such as multiple-line PSTN calls to expedite sending of multiple-address messages. One design has a 300-page memory plus store-and-forward capability. International character sets are available to give the PC-to-fax user more of a choice of formats when sending to Group 3 fax units.

Other arrangements for personal computing fax have the PC-fax functions in a PSTN communication server with optical disc archiving of all fax traffic. The *write once read many times* (WORM) optical disc has a complete unerasable record of all transmissions to Group 3 fax units.

3M Fax Xchange

This high capacity store-and-forward device has all of the features of a high-end PC-fax card and more. It can be accessed by up to one thousand users and has the capacity for three thousand fax transmissions per day with storage for up to two thousand pages. It can be interconnected to another 3M Fax Xchange by a packet switching network, using X.25 protocol. Group 3 fax units can send their documents into this switch where they are stored and forwarded to another Fax Xchange switch in high-speed, 64- or 56-kb/s error-free mode. The second switch sends in Group 3 mode to one or many destinations in the local area.

Protocol and speed conversion is possible with Group 3 from Group 1 and Group 2 fax units. A separate communication port provides for conversion of ASCII data to fax when receiving from a computer terminal, mainframe, or PC. Each of the seven PSTN ports must be configured for either receiving or sending. Simultaneous transmission is thus possible over six PSTN ports.

Some private branch exchange (PBX) switchboard suppliers furnish a service similar to a PC-fax card. Work stations can access modems in the PBX to send and receive files by Group 3.

Chapter 3
Group 4 Facsimile

3.1 THE GROUP 4 STANDARD

The market for Group 4 is quite different from the stand-alone, person-to-person communication that Group 3 now covers so well. Group 4 is better suited to computer-controlled network communication with multiaddress, store-and-forward, and electronic mail (message) systems.

Although the CCITT adopted recommendations for Group 4 in 1984, work continues to complete the standardization. The recommendation specifies the OSI 7-layer model, but none of the US networks exactly match its requirements. Full compliance is planned in the future.

There are digital communication services that offer Group 4 compatibility now, but worldwide connection to other parties is very limited. Operation of Group 4 machines on the regular telephone network (PSTN) is possible, but not in accordance with the Group 4 standards. The CCITT is still in the process of selecting a modem for the PSTN.

For the present, Group 4 will be limited to very specialized applications. Group 4 facsimile will not even start to replace Group 3 units for the next three years. The regular telephone network is slowly being converted to an all-digital network that eventually may be a part of the ISDN. When the PSTN becomes the ISDN, Group 4 will have the same worldwide network availability that Group 3 enjoys now.

Group 3 and Group 4 use the same techniques to process the picture signal before converting it to digital. The analog video output from the scanner is sampled and converted to a binary digital (black-white) signal. This is done by either a simple fixed-level slicing circuit or one that dynamically senses the background level. (See Chapter 4 for details.) The Group 4 facsimile scanner and printer must be capable of storing at least one page of compressed data. Meeting this requirement should cause no problem because existing designs typically store 60 pages or more. Groups 1, 2, and 3 fax units did not include the traditional DTE/DCE interface because the modem was built in. For Group 4 apparatus, a standard physical digital interface provides connection to the packet and circuit services of public data networks. In the future, the ISDN will be used for digital communication. For operation over the analog telephone network, the CCITT is defining a new high-speed PSTN modem for Group 4 fax and other telematic services. Proposed characteristics include:

- Asymmetrical duplex operation;
- Forward rate up to 14,400 b/s and possibly 16,800 b/s;
- Backward rates up to 450 b/s;
- Rapid data-rate determination at initiation and during the call;
- Start of data transmission 1.6 seconds after call establishment;
- V.33 trellis coding.

Computer Communications With Group 4

It is anticipated that Group 4 facsimile will usher in a new era in the facsimile business when ISDN becomes available everywhere. In addition to fax transmission to another user as in Group 3 facsimile, Group 4 will be an input-output device for a remote computer. At the present time there is a limited degree of facsimile-computer interaction utilizing PC-fax on the PSTN.

A few computer applications for Group 4 facsimile are:

Electronic Mail— The public data network can be accessed for the exchange of business mail.

Storage and Retrieval— Documents can be stored in a computer with availability for printing on a remote Group 4 fax unit.

Integration of Text with Graphics— Graphic data can be entered into a computer by a Group 4 scanner for use in a graphics and word-processing system. The operator will have the ability to add textual information or change the actual graphic image. The document could then be stored and distributed to Group 4 printers.

Optical Character Recognition (OCR) and Analysis of Graphic Data— Raw symbol and character data could be scanned by a Group 4 device and stored in a computer system, where pattern recognition could be performed.

3.1.1 The Seven-Layer OSI Model

The International Organization for Standardization (ISO) developed the seven-layer reference model of Open Systems Interconnection (OSI). The overall objective of the OSI is to define standards allowing systems to communicate. Existing standards are retained where possible, as new standards are developed. Subcommittee 16 (on OSI interconnection) of Committee 97 started the standard-making project in 1977. The reference model became the international standard IS 7498 in October 1984.

OSI provides the framework to ensure that all communication standards developed are compatible. A system which obeys applicable OSI standards in its communication with other systems is termed an "open system," often referred to as a "distributed system" in the United States. The OSI open systems concept is based on the need for an application process such as a Group 4 fax or other device anywhere in the world to interact with any application process at another location. The protocols will make possible worldwide linkage of computers and terminals from different vendors, electronic-mail systems, Group 4 fax units, and other application processes. As a result, competitive products and lower prices are expected. Users will have a choice of vendors and a variety of communication equipment: Manufacturers and vendors will have improved productivity and a broader market base.

Communication between Group 4 fax units or with other devices (application processes) is broken down into seven subtasks called *layers*. Each layer groups certain related aspects of data communication. (See Figure 3.1.1.1.)

A logical order was required for the continuum of functions involved, leading to the layering technique. Too many layers would make the structure overly complex, while too few would lead to the development of very complicated protocols. The decision on the structure was selected to be acceptable to the widest range of interests, both technically and politically. A seven-layer structure was developed and gained worldwide acceptance as the basis for the OSI architecture.

Each layer can be thought of as an envelope with the written information on the outside indicating where and how to deliver the contents. The actual data transmitted are in the innermost envelope. The seven envelopes used at the transmitting end are stripped off one at a time for each layer of the receiving end. The control information on the outside of each envelope is nondata information called *overhead*. This must be sent in addition to the data. For packet switched networks, this overhead is quite high. Therefore, facsimile is not well suited to transmission over packet switched networks.

Architecture is still being developed for a complete family of interfaces and protocols for worldwide distributed information and telecommunication systems.

The following is an example of the type of service that may be provided under OSI. After initially programming the Group 4 fax unit, a letter could be typed with a "copies to" list. The fax unit looks up the calling numbers stored in its memory for the names on the "copies to" list and automatically sends to all of their offices. If programmed into the receiving fax unit, the letter would be delivered to any other location in the world where the recipient happened to be.

In the United States, OSI recently received a boost. As of 1990, all federal government purchasing of networks must conform to OSI standards, as specified in the Government Open Systems Interconnection Profile (GOSIP). Some analysts believe the federal purchasing mandate and the strong support for the model in Europe will make OSI networks commonplace by 1995.

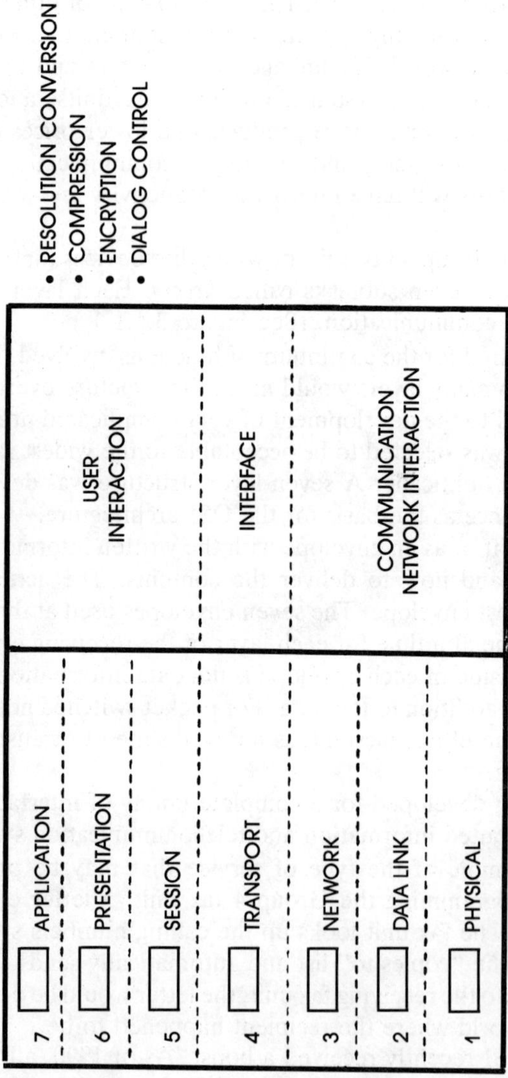

Figure 3.1.1.1 The OSI model.

Modifications of the original Reference Model reflect newservices such as:

- Connectionless-Mode Transmission;
- Security Architecture;
- Naming and Addressing;
- Management Framework.

ISO specification IS 7498 is identical to the CCITT specification X.200 reference model of OSI developed by Study Group VII. Figure 3.1.1.2 shows some of the standards that are applicable to Group 4 facsimile and how they interface with other telematic services.

Layer 7 — Application

The highest level, the application layer, is the user interface between Group 4 fax or other service and the OSI environment. The application process can be manual, such as operation of an automated banking terminal, a computerized process where a program running in a computer center accesses a remote database, or a physical process whereby a robotic arm is controlled remotely by a computer program.

A great amount of work is continuing on the following application layer standards:

- T.563 "Terminal Characteristics for Group 4 Facsimile Apparatus";
- T.521 "Communications Application Profile BT0 for Document Bulk Transfer Using the Session Service Defined in Recommendation T.62bis";
- T.503 "A Document Application Profile for the Interchange of Group 4 Facsimile Documents";
- Recommendation T.6 "Facsimile Coding Schemes and Coding Control Functions for Group 4 Facsimile Apparatus."

Recommendations T.521, T.503, and T.563 appear in the new CCITT Blue Book as replacements for Recommendation T.5, "General Aspects of Group 4 Facsimile Apparatus," as it is defined in the CCITT Red Book.

Layer 6 — Presentation

The presentation layer protocol handles session establishment and termination requests, and it preserves the meaning of the data while resolving syntax differences. The formats are selected for data and control signals. The most important standards currently in place are:

- X.226 "Presentation Protocol Specification for Open Systems Interconnection for CCITT Applications";

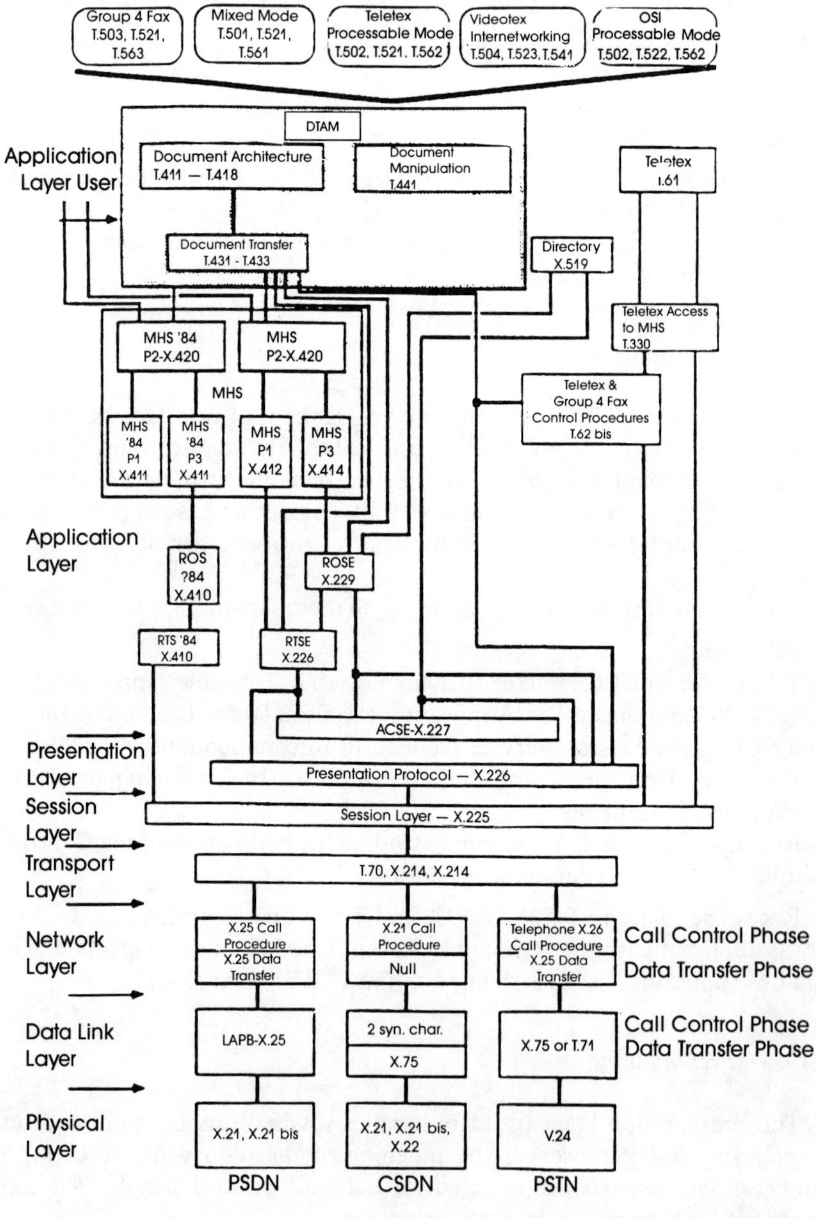

Figure 3.1.1.2 Group 4 facsimile recommendations in the OSI model.

- X.208 "Abstract Syntax Notation One (ASN.1)";
- X.209 "Basic Encoding Rules for ASN.1."

The ISO has counterparts to these standards. Recommendation X.226 specifies the procedure for transfer of data and control information from the presentation layer of the sending system to that of the receiving system.

Recommendation X.208 is used as a semiformal tool to define protocols and a mechanism for the transfer of user data. A number of simple types (i.e., integer, character, real) are defined. These simple types are then used to form more complicated ones.

Recommendation X.209 is a set of basic encoding rules used to derive the specification of a transfer syntax for values of types that have been defined by using X.208.

Layer 5 — Session

The session layer protocol establishes, manages, and releases the communication connection. This layer provides functions such as the start and end of document transmission, acknowledgement of pages received, and error recovery if abnormal conditions are encountered. Documents may be transmitted in one direction or in alternate two-way exchange within the same session. Session protocol functions as a mediator between application processes. It synchronizes their dialogue by binding and unbinding the application processes into a communicating relationship and helps to manage their exchange of data. The session layer combines different applications into a common language for calling another application process.

Recommendations of importance to the session layer are:

- X.215 "Session Service Definition for Open Systems Interconnection for CCITT Applications." X.215 defines the services provided to the presentation layer by the session layer.
- T.62bis "Control Procedures for Telex and Group 4 Facsimile Services Based on Recommendations X.215/X.225." T.62bis defines the set of rules for using the OSI session services and the additional requirements to conform to the control procedures for both teletex and Group 4 facsimile services.

Layer 4 — Transport

The transport layer provides a communication path to other desired functions as a bridge between the application-related functions and the transmission-related functions. Additionally, it provides a consistent transport service in association with the lower three layers.

Functions performed by this layer are:

- Error Recovery;
- Error Detection;
- Flow Control;
- Multiplexing.

Recommendations of importance to the transport layer include:

- T.70 "Network-Independent Basic Transport Service for the Telematic Services." T.70 defines the network-independent basic transport service applicable to both teletex and Group 4 facsimile terminals connected to the CSDN, PSDN, and PSTN. Transport layer procedures for ISDN are subject to further study.
- X.214 "Transport Service Definition for Open Systems Interconnection (OSI) for CCITT Applications." X.214 defines the services provided by the transport layer to the session layer at their boundary.
- X.224 "Transport Protocol Specification for Open Systems Interconnection for CCITT Applications." X.224 defines a common encoding and different classes of transport protocol procedures to be used with various networks and their quality of service. In addition, mechanisms are defined to enhance the quality of service as it relates to throughput rates, error rates, integrity of data requirements, and reliability requirements.

Layer 3 — Network

Routing and relay functions through switched telecommunication media are the primary functions of the network layer. The means to establish, maintain, and terminate the network connections between systems is also under the network layer's control. Network connection is transparent to the transport entities, ensuring that service provided at each end of the network connection will be the same. This holds true even if the end-to-end connection spans several subnetworks.

Standards of importance to the network layer include:

- X.25 "Interface Between Data Terminal Equipment (DTE) and Data Circuit Terminating Equipment (DCE) for Terminals Operating in the Packet Mode and Connected to Public Data Networks by Dedicated Circuit." X.25 defines mechanical, electrical, functional, and procedural characteristics to activate and deactivate the physical links between the DTE and DCE. In addition, it defines the packet layer procedures for the exchange of control information as well as for user data at the DTE-DCE interface.
- X.21 "Interface Between Data Terminal Equipment (DTE) and Data Circuit–Terminating Equipment (DCE) for Synchronous Operation on Public Data Networks." X.21 defines both the physical characteristics and call con-

trol procedures for a general-purpose interface between DTE and DCE employing synchronous transmission operating on public data networks.

Layer 2 — Data Link

The major function of the data link layer is reliable transfer of all information over the physical transmission media. The maintenance and release of data-link connections among all network entities is performed within this layer. Additionally, it is responsible for detecting and correcting errors that may occur in the physical layer and enabling the network layer to control the interconnection of data circuits within the physical layer.

Recommendations of importance to this layer include:

- X.75 "Terminal and Transit Call Control Procedures and Data Transfer System on International Circuits Between Packet-Switched Data Networks." X.75 defines the physical, data-link, and network-layer protocols used for interworking between two packet switched public data networks.
- T.71 "LAPB Extended for Half-Duplex Physical Level Facility." T.71 defines the LAPB protocols at the data link layer for DTES operating in the packet mode to access a public data network or for two directly connected packet mode DTEs to communicate.

Layer 1 — Physical

The physical layer deals with the transmission of a bit stream, regardless of its meaning, across a physical communication medium. This medium may be copper wires, optical fibers, or a radio link. The physical layer provides mechanical, electrical, functional, and procedural characteristics to activate, maintain, and deactivate physical connections to the transmission media. Connections may be either point-to-point or multipoint.

Recommendations of importance to this layer are:

- X.21bis "Use on Public Data Networks of Data Terminal Equipment (DTE) which Is Designed for Interfacing to Synchronous V-Series Modems." X.21bis defines the physical characteristics and call control procedures for a DTE which operates with synchronous V-series modems to access a public data network.
- V.24 "List of Definitions for Interchange Circuits Between Data Terminal Equipment and Data Circuit–Terminating Equipment." V.24 applies to the interchange circuits at the interface between DTE and DCE for the transfer

of binary data, control and timing signals, and analog signals. This recommendation applies to both synchronous and asynchronous data communication, data transmission on leased line service, and data transmission on switched network service.

3.1.2 Resolution

Resolution is the primary factor affecting quality of the output document as well as the number of bits required for transmission. Four resolutions were chosen for Group 4. Each has a horizontal and vertical tolerance of ± 1%.

The 200 × 200 resolution was selected because it is the current high resolution option for the Group 3 system. Most Group 3 systems manufactured today have this capability so its inclusion in the Group 4 system should help contribute to compatibility between Group 4 and Group 3 systems.

The 300 × 300 resolution is better for small typefaces and provides a higher quality of output images. The primary rationale for this resolution is that the human observer, even without the aid of optical magnification, can perceive more detail than is provided with the lower resolution. To the casual observer, the received document could pass as an original. This resolution matches the 300-dpi resolution of the laser printers generally used with personal computers, making their use as Group 4 components relatively easy. Teletex compatibility requirements can also be satisfied with 300 lines/inch.

The 400 × 400 resolution is required for good reproduction of fine detail, such as Japanese *kanji* characters. Group 4 has very high quality output images, providing a new and distinct service as compared to Group 3. In a 400-lines/inch image there are 78% more pels than in a 300-lines/inch image. Only ⅓ more bits are required to send the compressed image, however, since the compression ratio at 400 lines/inch is much greater. If the 400-lines/inch detail is needed, the relatively small increase in cost for communication and computer storage can usually be tolerated.

3.1.3 Three Classes of Group 4 Facsimile Apparatus

Teletex and Group 3 fax recommendations were adopted by the CCITT in 1980. Group 3 uses the public switched telephone network, while teletex uses public data networks. When Group 4 fax specifications began, the CCITT insisted that the protocols developed for teletex be used for Group 4, and that the two services would communicate with each other. When Group 4 fax received teletex signals, it would decode the ASCII-like symbols and assemble them into a printing line in a manner similar to the 240-dots/inch teletex units. When sending from

Group 4 fax units to teletex units, areas of the page containing symbols would be teletex coded for higher transmission efficiency.* Teletex supporters did not plan to use bit-mapped graphics with Group 4 coding. Because the fax supporters knew that most Group 4 users would not pay extra for teletex, recommendations were adopted for three classes of Group 4 fax units. (See Figure 3.1.3.1.)

CLASS	1	2	3
Pel-density of scanner-printer (pels/25.4 mm)	200	300	300
Pel transmission density (pels/25.4 mm)	200	300/200	300/200
Pel transmission conversion capability	not required	yes	yes
Mixed-mode capability	not required	not required	yes
Optional pel density of scanner-printer	300/400	400	400
Combined with pel transmission density (pel/25.4 mm)	400/300/200	400/300/200	400/300/200
Storage	not required	not required	yes

Figure 3.1.3.1 Characteristics of three classes of Group 4 fax units.

- Class 1 fax units send and receive facsimile coded documents. The only resolution required is 200 pels/inch (25.4 mm), but Class 1 fax units available today also have the higher optional resolutions of 300 and 400 pels/inch. Recommendations T.6, T.503, and T.521 apply.
- Class 2 fax units send and receive facsimile coded documents with Class 1 capabilities, adding 300 pels/inch as a mandatory requirement. In addition,

* Only 8 bits per character are needed to send teletex, compared with about 90 to 180 bits to send the same information in Group 4 (8 × 10 inches page area). Teletex requires 80 characters/line × 60 lines/page × 8 bits/character = 2880 bits. Group 4 at 200 lines/inch, requires 320,000 bits for the same area. Assuming a compression efficiency of 10:1, 32,000 bits are needed; 32000/2880 = 11.1. Group 4 at 400 lines takes only twice as many bits due to increase in compression efficiency.

Class 2 can receive teletex and mixed mode documents. Recommendations T.6, T.503, T.521, T.60, T.61, and T.561 apply.

- Class 3 fax units send and receive facsimile coded documents with Class 1 and Class 2 capabilities. In addition, Class 3 can generate and send teletex and mixed mode documents. Recommendations T.6, T.503, T.521, T.60, T.61, and T.561 apply.

Three classes of Group 4 fax units are defined within the standard, but only Class 1 has a real chance for widespread implementation as Classes 2 and 3 may be eliminated.

3.1.4 Basic and Optional Coding Schemes

Recommendation T.6, "Facsimile Coding Schemes and Coding Control Functions for Group 4 Facsimile Apparatus," details the coding schemes. The basic facsimile coding scheme for Group 4 apparatus is a two-dimensional code, which is very similar to that used for Group 3 facsimile. Optional coding schemes for both gray-scale and color images have been left for further study.

The Group 4 coding scheme, often called modified modified read, is a higher efficiency version of the Group 3 modified read code and uses the same encoding tables. See Chapter 3 for a description of the various coding modes (pass, vertical, and horizontal). The position of each changing pel on the current coding line is encoded with respect to the position of a corresponding reference element situated on either the coding line or reference line that directly precedes the current coding line. After the current coding line has been encoded, it will become the reference line for the line directly beneath it. When coding the first line in a document, the reference line is an imaginary white line.

The reader will note that the code used for Group 4 has the following differences from the Group 3 code: the K-factor is equal to infinity, the minimum scan line time is equal to zero, and no end-of-line codes are used.

The transport service for Group 4 facsimile can be provided by several types of network. These include the circuit switched public data network (CSPDN), packet switched public data network (PSPDN), public switched telephone network (PSTN), and eventually the ISDN. The Group 4 facsimile apparatus will provide for automatic answering, transmission, reception, and clearing, regardless of the type of network used.

3.2 NETWORK-RELATED REQUIREMENTS

As the establishment of public networks for data transmission and ISDNs for integrated services grew, a need was seen for the development of standard "user classes of service." The user class of service details data transmission service with standardized data signaling rate, call control signaling rate, and data terminal equipment operation mode.

3.2.1 Public Switched Telephone Network

One advantage of the PSTN over other networks is its universal availability. The PSTN functions as the backbone of Group 3 facsimile apparatus and is viewed as a viable network for Group 4. One of the major disadvantages is that the system was designed to handle voice, not digital data. The rate at which data can be reliably transmitted is lower than over other types of network. The functional and procedural aspects of the interface have yet to be developed. Additionally, the modulation-demodulation schemes and the bit rates are still to be defined. It is widely believed that if the Group 4 facsimile terminals are to transmit over the PSTN, they will require modems somewhat similar to those currently used for Group 3, but with higher speeds and improved performance.

3.2.2 Circuit Switched Public Data Networks

The functional and procedural aspects of the interface are defined by CCITT Recommendation X.21. Bit rates are specified in Recommendation X.1, "International User Classes of Service in Public Data Networks and Integrated Services Digital Networks (ISDNs)." Group 4 fax has bit rates of 2.4, 4.8, 9.6, 48, and 64 kb/s for user classes of service 4 to 7 and 19. The signals in the call control phase are required to use International Alphabet No. 5.

3.2.3 Packet Switched Public Data Networks

The functional and procedural aspects of the interface are defined by CCITT Recommendation X.25, which specifies that OSI layers 1, 2, and 3 are used for transmission over the PSPDN.

Recommendation X.1 specifies data signaling bit rates of 2.4, 4.8, 9.6, 48, and 64 kb/s, conforming to user classes of service 8 to 13. Because the signaling rates of 9.6 and below are the same as the Group 3 standard, compatibility between the Group 4 and Group 3 apparatus should be easier to achieve.

3.2.4 Integrated Services Digital Network (ISDN)

The ISDN is an internationally standardized extension of the ongoing conversion of the regular telephone network (PSTN) from analog to digital. The ISDN will provide an all-digital switched channel, end-to-end. The process started about

20 years ago when short-haul analog multiplexing systems like N-carrier began to have problems keeping up with the demand for telephone circuits. The quality of transmission had also degraded. The T-1 digital system was then introduced into the telephone plant in the Los Angeles area to handle regular voice calls. Two pairs of copper wires carry a 1.544-Mb/s signal for 24 voice channels. Today, telephone company central offices in most metropolitan areas use digital switching.

Instead of using 300 to 3000 Hz analog channels between a residential or small office customer and the central office, ISDN will provide 144 kb/s. Two full-duplex 64-kb/s channels plus one full-duplex 16-kb/s channel are planned.

ISDN is well suited to Group 4 fax communication. Existing Group 4 fax units must slow down to 48 or 56 kb/s to use existing digital channels. These channels are expensive, costing about $1000 per month as a connection fee and an additional charge for each kilobit sent. Expensive lines are also needed to reach the telephone company's central office. Initially, charges for operation of fax units on the ISDN will probably be higher than existing PSTN channels.

What does this mean for the future of Group 3 fax? By using a voice telephone type codec, Group 3 fax can continue to be switched to worldwide destinations in the same way as voice. A 64 kb/s Group 3 interface would allow much higher speeds.

Field trials of ISDN in phase 1 have been operating in isolated pockets for some time. Not until 1995, however, will ISDN have replaced most of the existing telephone service and begun to offer a universal high-speed service for fax. Meanwhile, broadband ISDN is rapidly developing and will offer much higher speed channels.

3.3 MIXED MODE

In a mixed-mode system, the information on a page is divided into two parts, the first being symbols (letters, numerals, punctuation) and the second being graphics (logos, signatures, sketches). Mixed mode must support the service requirements for both teletex and Group 4 facsimile. Recommendation T.72 in the CCITT Red Book has been replaced by the following three recommendations:

- T.561 "Terminal Characteristics for Mixed Mode of Operation MM1";
- T.501 "An Application Profile MM1 for the Interchange of Formatted Mixed Mode Documents";
- T.521 "A Transfer Application Profile BT0 for Formatted Document Bulk Transfer."

By splitting the mixed-mode operation into separate recommendations, it will be easier to keep them in alignment with the work currently being done to define the T.400 series of recommendations for Document Transfer Access and Manipulation (DTAM).

Recommendation T.561 defines the general characteristics of the equipment that supports mixed-mode operation. Basic features for mixed-mode operation include the ability to create, transmit, and receive documents that conform to the document application profile, as specified in Recommendation T.501; the ability to handle the ISO A4 paper size; and the ability to handle characters and raster graphics, as defined in Recommendation T.501.

Recommendation T.501 allows mixed-mode documents to be interchanged only in a specified format, which will provide the receiver with the opportunity to reproduce the document as intended by the sender. The document application profile has been designed to support both character content as used by a word processor and raster graphics as used by Group 4 facsimile.

Recommendation T.521 defines the communication application profile for the bulk transfer of formatted documents. The recommendation is based on the session services as defined in Recommendation T.62bis.

Chapter 4
Scanning Technology

The scanner is the part of a fax unit that converts the marks on the page being sent into electrical signals. A photosensor (electric eye) looks at a very small spot (pel), determines whether it is black or white, and generates a strong or weak electrical pulse. To read a page electronically, all of the spots in a thin strip 1/100th inch high across the top of the page are read, one at a time, starting with the upper left-hand corner. A stepper motor moves the page down and the photosensor reads the next strip below. Successive strips are read until the whole page is converted into electrical pulses. While early scanners used only one photosensor, Group 3 fax units use a strip (array) of 1728 photosensors, one for each spot across the page. An electronic clock, which ticks about a million times a second, causes one photosensor to be read per tick. The whole page can be read in 10 seconds or less.

Both the CCD chip and contact image sensor types are available in 1728, 2048, or 2432 photosensor arrays.

Self-contained scanners are available for personal computers and can be used with PC-fax cards, *optical character recognition* software, or for entering imagery from hard copy into computer memory for desk top publishing or other programs. When scanners first became available, most were based on facsimile scanning technology with 203 dpi. Now, most scanners for PCs offer higher resolutions of 300 or 400 dpi.

4.1 CCD CHIP SCANNERS

The camera type of optical imaging system has a small *charge coupled device* (CCD) photosensor chip where the film would be in a camera. There are 1728 photosensors in a straight line across the chip. A reduced-size image of the page is focused onto the CCD chip in a manner similar to that of a 35-mm camera. (See Figure 4.1.1.)

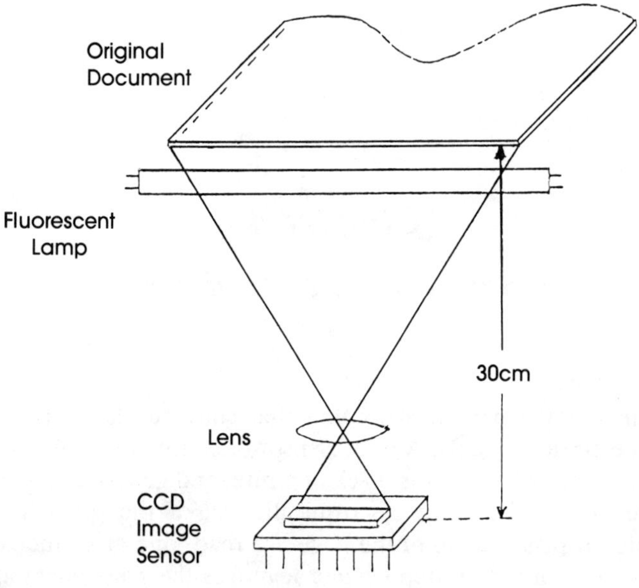

Figure 4.1.1 Camera type of CCD scanner.

The scanning line position on the page is determined by the part of the image falling onto the photosensitive strip of the chip. A fluorescent lamp near the page being sent illuminates the scanning line. As with a camera, the lens cannot be very close to the page. A distance of at least 12 inches is required between the scanning line and the CCD chip. This distance controls the minimum size of the fax unit. To reduce the size of the fax unit, two or three mirrors may be placed to fold the optical path into a more compact space. Precision optical alignment is then required with a rigid mounting for the optical components. Limitations on lens designs result in slight geometric and other distortions. The 8½-inch page width matches that of the 10:1 reduced-size image on the CCD chip.

Each photosensor element across the CCD chip is a light-sensitive resistor connected in parallel with a small capacitor. The capacitor starts with a full voltage charge. If the image spot is white, the photosensor has a very low resistance and the voltage drains off the capacitor. A black image spot causes the photosensor to have high resistance and the capacitor is not discharged. The voltage at each photosensor element, when read, depends on the black-white image pattern for the thin-strip scanning line across the page being sent. The picture signal is the current pulse into the capacitor as the photosensor is charged for the next scanning line. Each capacitor has the whole time between scanning lines for the voltage to leak off, providing good sensitivity.

A clock and shift register starts sampling at the left edge photosensor with the current pulse amplitude representing brightness of the imaged spot. The amplitude of this video pulse is highest for white and lowest for black. Gray pixels have intermediate amplitudes. The next clock pulse connects the adjacent sensor to the output. The process repeats until the last sensor for the right side of the page is sampled. After this 1728th pixel is sampled, a stepper motor moves the page down 0.01 inch, and clock sampling continues across the next scanning line.

Each sensor is a pixel whose picture signal is connected in sequence to an *automatic background control* (ABC) circuit. The ABC corrects for pages with a background that is not fully white. For a gray or colored background, the gain of the ABC increases until its output is the same voltage as for a white background.

An electronic shading circuit then corrects for fall-off in brightness near the edge of the page and variations in sensitivity of the individual photosensor elements. This circuit stores in a RAM the white signal voltage for each pixel across the page when a white border is scanned. Correction is then made for each pixel sensitivity while the page is being sent.

The voltage of the corrected picture signal at this point varies directly with the darkness of marks being scanned. This signal is then changed to black or white only by an analog-to-digital (A/D) converter. The A/D converter has an input threshold voltage that separates the mark voltages from those of the background. The output of the A/D is a digital signal with each pixel converted to a one-bit (black or white) pel. This step is necessary for efficient coding of the picture signal. The black-white effect on the page being sent is the same as that produced by an office copier.

4.2 CONTACT IMAGE SENSOR SCANNERS

Use of a CIS scanner instead of the camera type of scanner described above gives a major size reduction of Group 3 fax units. CIS uses a strip of photosensors the 8½-inch width of the page being scanned instead of about 0.8 inch across a CCD chip. All of the scanner elements are contained in a small bar mounted so as to be in contact with the page being scanned. The bar contains lamps for illumination of the scan line, optics if any, the sensor array, and video signal processing circuitry.

The photosensor strips usually are three silicon CCD or cadmium sulfide–cadmium selenide strips, each three inches wide, or a single full-width strip of amorphous silicon. The three strips are assembled end to end for an 8½-inch wide scanner. The strips may be staggered with overlap to allow a mounting area for the sensors at the end of each strip. Analog memories fabricated with the CCD chips compensate for the offset stagger required between adjacent strips. Signals from the strips that first see the scan line being read delay their outputs for a few paper-feed steps until signal for the remaining sections of the same scan line is available from the other strips. One company, Matsushita, has developed a manufacturing technique for mounting all CCD chips in a straight line, thus avoiding the stagger compensation problem.

Amorphous silicon arrays use a coating process that makes a single array the width of the whole scanning line. Amorphous silicon is less expensive, but the signal-to-noise ratio is less than CCD types.

CIS scanners have the advantage of the exact dot position across the page and the same sharpness of focus on the end of a line as in the center.

The discussion of operation and circuitry given in Section 4.1 above also applies to CIS scanners. Figure 4.2.1 shows a compact CIS scanner design.

Either optics or full-contact types of imaging methods are used for CIS scanners.

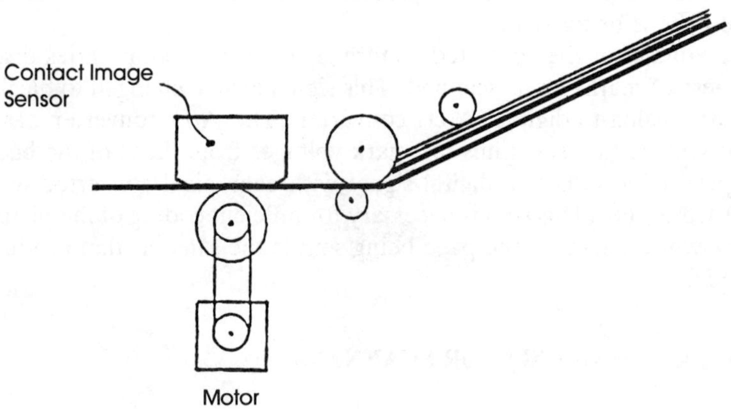

Figure 4.2.1 Stationary head CIS scanner.

4.2.1 Optics CIS

A narrow (8½-inch) bar containing rod-lens (Selfoc) fiber optics is mounted between the page being scanned and the photosensors. Only about five centimeters are required between the sensor array and the page being scanned. (See Figure 4.2.1.1.)

A fluorescent lamp, mounted close to the scanning line, illuminates the page being scanned. The scanning line is directly above the fiber optics rod-lens array. It is focused with a 1:1 magnification onto the long image sensor strip below it. Each photosensor element is the same size as the pixel image that it sees.

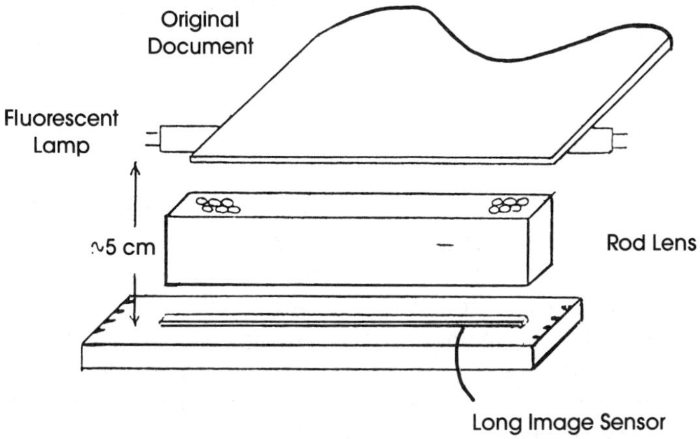

Figure 4.2.1.1 Contact image optics type of scanner.

In an alternative arrangement, illumination of the line being scanned is provided by one or two rows of *light emitting diodes* (LEDs). The LED arrays have a built-in lens on the end of each diode. (See Figure 4.2.1.2.)

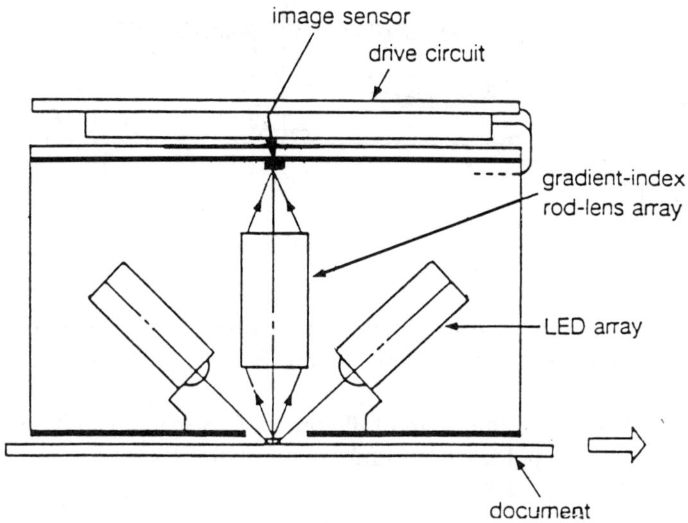

Figure 4.2.1.2 Contact image scanner.

The optics CIS unit is also available in a moving scan-head design where the page to be sent is held face down on a glass plate and the scan head reads through the glass as it moves down the page.

4.2.2 Full-Contact CIS

The full-contact type of CIS is constructed with a hole through the center of each small photosensor element. The LED array for illumination of the scan line is on one side of the photosensor strip and the page being scanned is on the other. Light from a LED passes through a photosensor hole (window) to illuminate a very small spot on the page being scanned. A transparent coating on the page side holds the paper a small distance from the sensitive side of the photosensor strip, allowing light from the illuminated spot to reach the sensitive area of the photosensor surrounding this hole. (See Figure 4.2.2.1.)

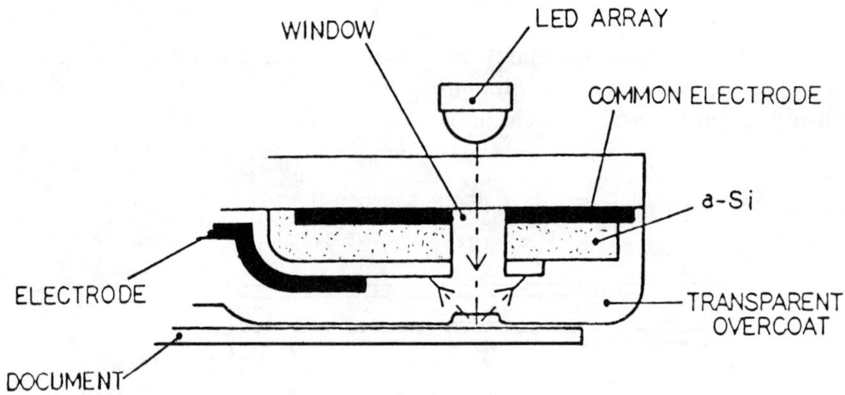

Figure 4.2.2.1 Full contact CIS cross section.

Practically none of the light reflected by one spot is picked up by adjacent sensors. The optical efficiency of this system is higher than with rod-lens optical imaging.

Higher resolution camera and full-contact CIS scanners with 300 or 400 dpi are available for Group 4 fax units and for use with PCs.

Chapter 5
Printing Technology

Printing technology is a critical part of the quality of the output copy produced by a facsimile system. Group 3 fax units use direct recording on thermal-sensitive paper, thermal transfer recording onto white paper, or xerographic recording on white paper.

5.1 THERMAL PAPER FAX RECORDING

Most fax receivers today record directly onto heat-sensitive paper with a thermal print head of 1728 wires in a row across the 8½-inch paper width. Wider heads use 2048 or 2432 wires.

5.1.1 Thermal Print Head Operation

The print head has a row of very small resistor-element spots across the recording paper width, 1728 for 8½ inches or 2048 for 10 inches. The thermal recording paper touches this row of resistors. A pulse of current through a resistor causes it to become hot enough to mark the paper in a spot about 0.005 inch in diameter. The spot temperature must be changed from nonmarking temperature to marking temperature and return to nonmarking before the paper steps to the next recording line. The recording cycle may be 2.5 to 5 ms. In one design, 24 V at a power level of 0.5 W is applied to a marking resistor for about 0.6 ms. Marking temperature is about 200° F. Other head designs use 12 or 6 V.

Either thin-film or thick-film technology is used for making the print heads. Thin-film heads take lower recording power and can be made with higher resolution. The head wires and integrated circuit elements are made in one compact head assembly. (See Figure 5.1.1.1.)

Thick-film print heads have good immunity to scratches and abrasive wear and enough resolution for Group 3 fax units.

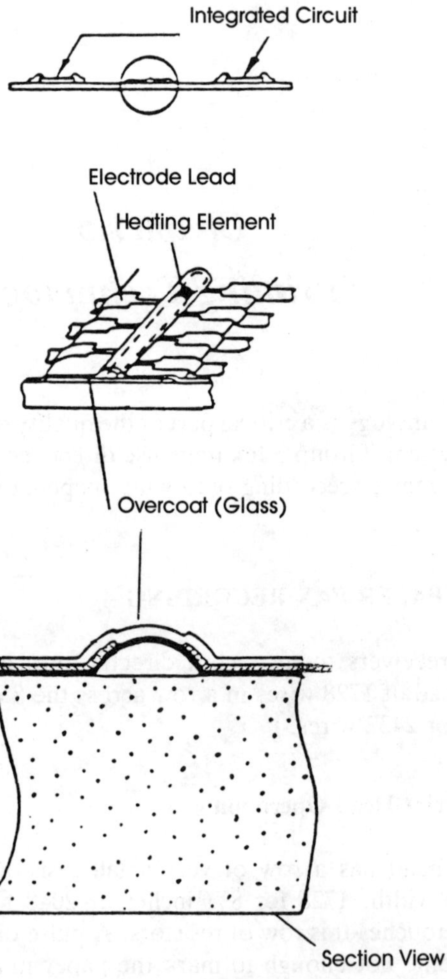

Figure 5.1.1.1 Thin-film thermal print head.

5.1.2 Pros and Cons of Thermal Fax

Although most customers prefer plain paper recording, there are many factors which make them accept direct thermal paper recording.

Operator involvement is minimal— The only regular chore in using the fax equipment is changing the roll of recording paper. This is usually a very simple task like opening a cover and dropping in the paper roll. There are no toner powders to be changed or to cause black dirt problems.

Reliability is extremely high— The print heads normally last the life of the fax unit without service calls or user maintenance.

Cost is low— The lowest cost fax units use thermal printing. Thermal paper cost is reasonable, usually less than five cents per page. See Section 9.3.2 for caution on buying paper. To prevent gumming up of the print head, be sure the paper you buy is approved by the fax machine manufacturer.

Size is small— They use an integrated print head design, where the print driver electronics is in an integral package with the thermal wires in the print head.

Printing speed is adequate for Group 3 fax— A whole page may be printed in less than 10 seconds.

Print quality is good— The recorded copy has saturated black marks with sharp edges on a slightly gray background.

Storage could be a problem— Properly chosen paper type and storage conditions will give a five-year copy life. Contrast of recording deteriorates if thermal paper is exposed to light or heat. For archival records an office copier print may be used.

5.2 PLAIN PAPER FAX RECORDING

5.2.1 Thermal Transfer

Thermal-transfer fax units write on white paper instead of light gray thermal paper. Recorded copy looks and handles about the same as copy from an office copier. Storage is not restricted by light or heat conditions as thermal paper is. Thermal-transfer fax recorders use the same type of print head as thermal-fax recorders, but the marking is made on white paper from a black sheet similar to carbon paper. A roll of white paper and a roll of thermal-transfer film (often called ribbon) are used, with the black film placed between the thermal head and the white paper. (See Figure 5.2.1.1.)

The black thermal-transfer film moves with the paper at the same rate so that they are used in the same quantities. Heat from the hot spots across the recording line causes the black marks to be made on the white paper. Some of the newer fax units use a stack of cut sheet white bond paper instead of a roll of paper. The marking process is somewhat similar to making a carbon copy, but with heat substituted for pressure. Thermal-transfer recordings are much sharper than carbon copies, however.

The film rolled up on a take-up reel inside the fax unit is removed when the roll of white paper is replaced. It has a negative record of all the pages recorded on its companion roll of white paper and can be retained for backup purposes. Before discarding this roll, consider whether the information it contains should be kept confidential.

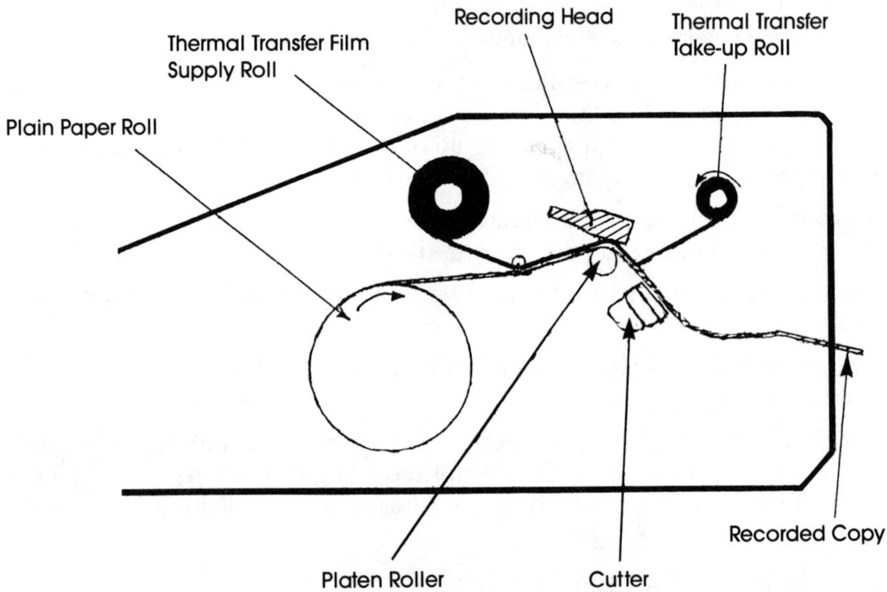

Figure 5.2.1.1 Thermal-transfer fax recording.

The cost of both the fax unit and recording supplies is somewhat higher than for thermal-paper recorders. Regular thermal paper can often be used as an alternative recording medium in thermal-transfer recorders. The take-up reel is a mechanical part that may contribute to a somewhat lower overall reliability of the fax unit.

Early thermal-transfer fax units required a very smooth (highly calendared), white recording paper, which did not look or feel like white bond paper. Much progress has been made in achieving the bond paper ideal. Originally, the black transfer ink would not mark in the "valleys" of bond paper, making a very spotty recording. The transferred ink has been made more viscous and fills in the rough spots for more even coverage of the entire paper marking area. Good-quality bond paper is required, but progress is being made in developing systems that work on much coarser papers.

5.2.2 Laser Fax Xerographic Recorder

Xerography, the original form of electrophotography, was invented by Chester F. Carlson on October 22, 1938. His U.S. Patent 2,221,776 was issued two years later. Laser xerographic printers operating at 300 dpi are a well established technology for office use in word processing, desk top publishing, and now PC-fax systems. Similar laser printers, adapted to work at 400 dpi, are used as the recorders in almost all Group 4 fax units. The quality of recorded copy is superior to that of thermal or the thermal-transfer types used in Group 3 fax units. The laser printer's size and price, however, have prevented it from making inroads into the Group 3 fax market.

Laser fax recorders are similar to office copiers, but the imaging is made by a fax signal that modulates a laser beam, rather than direct optical imaging of a page to be copied. (See Figure 5.2.2.1.)

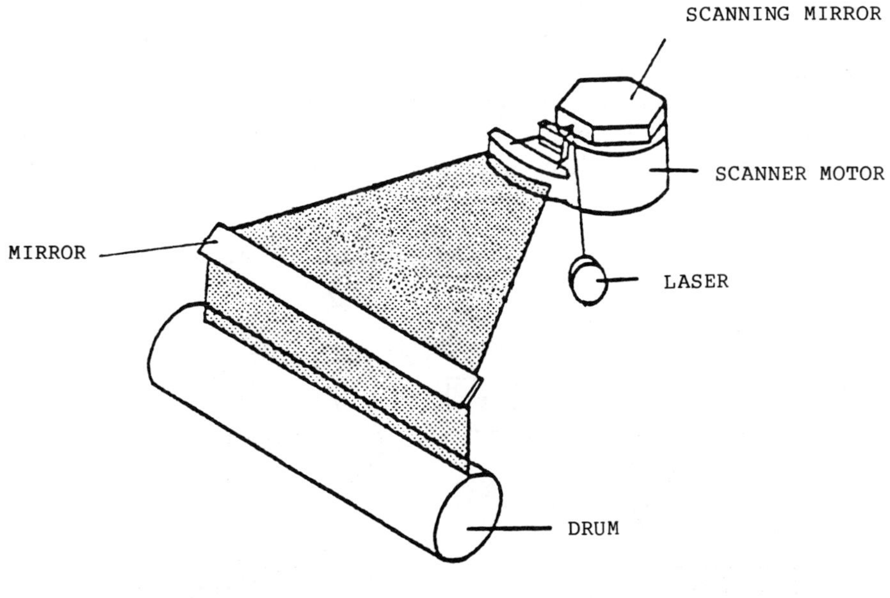

Figure 5.2.2.1 Laser fax xerographic recorder.

A small-diameter beam from a semiconductor laser is swept across a photosensitive drum by a rotating polygon mirror, making one recording line per mirror face. An imaging lens system keeps the spot in focus along the recording line. The semiconductor drum rotates the height of one recording line for each mirror-face sweep. Some designs may use a photosensitive belt instead of a drum.

5.2.3 LED Linear Array Recorder

A LED contact type of linear array can be designed into a xerographic recorder instead of a laser and rotating-mirror system. This simplifies the mechanical assembly of the recorder and increases its reliability by eliminating the rotating mirror. (See Figure 5.2.3.1.)

A LED print head with 300 or 400 dpi is focused onto the photosensitive drum with a fiber optics rod-lens array. The drum has been charged by a high voltage corona unit (electrostatic charge) before reaching the recording line. The charge is maintained in the dark, but is discharged at those points along the line exposed by light from the laser. The charge-pattern image remaining on the drum corresponds to the black marks on the page being sent by the fax transmitter. As this image passes the development box, black toner powder is picked up to reproduce a copy of the original in mirror image. A sheet of paper is fed onto the drum surface to start in step with the page image. As the drum rotates, the sheet passes the transfer-corona station and the toner image transfers from the drum to the paper. The image is heat or pressure fixed to the sheet to form a permanent copy of the page sent. Recording widths of 8½ inches and wider are available. The focus and spot size of the image are the same for all points. (See Figure 5.2.3.2.)

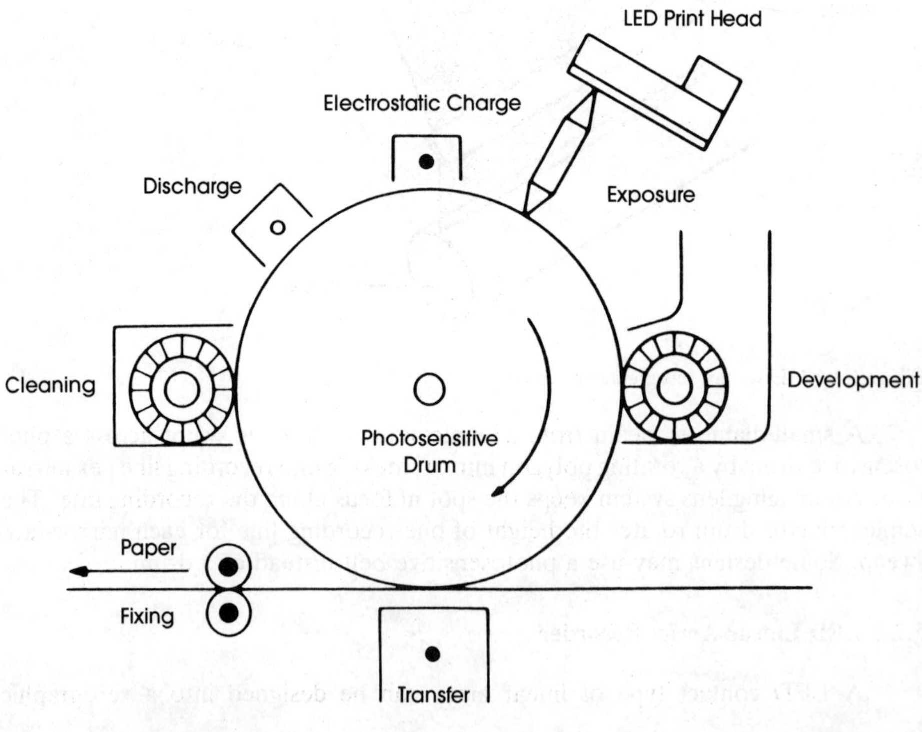

Figure 5.2.3.1 LED linear array recorder.

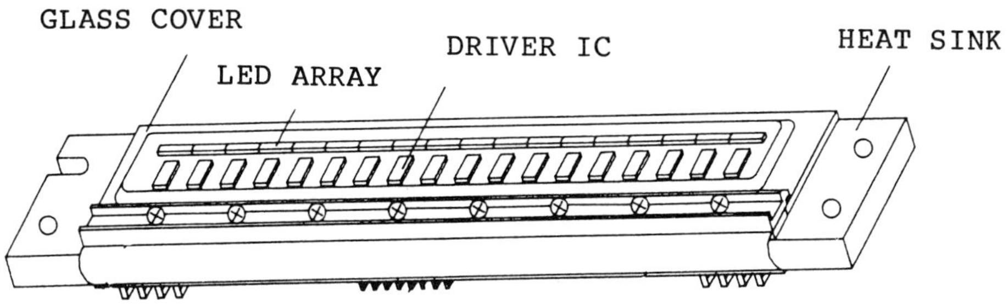

Figure 5.2.3.2 LED linear array print head.

5.2.4 Liquid Crystal Shutter (LCS) Printer

A *liquid crystal shutter* (LCS) printer performs like the 300-dpi LED printer described above. A linear array strip of light elements is the same length as the recording paper width. The liquid crystal elements selectively stop the light passing through them to write on the photosensitive drum. Very simple optics image the light spots on the photoconductive drum of the recorder.

This type of print engine could be substituted for a laser printer in Group 4 facsimile designs. A typical LCS printer matches the LED print quality and produces eight pages per minute.

Chapter 6
Advanced Coding of Binary Images

Both approximation algorithms and information-preserving algorithms have been developed. The information not restored in the approximation algorithms is set at a level where the received image is acceptable to the user, giving a higher compression factor and, thus, reduced sending time.

6.1 OVERVIEW

Figure 6.1.1 is a diagram showing the hierarchy of compression techniques for the efficient coding of black-white or binary images. There are two broad categories of compressiontechniques: *information-preserving* and *approximation* techniques.

Information-preserving techniques may be defined as being lossless or reversible in the sense that the output image will be a perfect replica of the input image if there are no transmission errors. There are two classes of compression techniques that preserve information: *nonadaptive* and *adaptive.* The nonadaptive techniques employ a coding algorithm that is independent of the statistics of the particular image being transmitted, whereas adaptive techniques employ an algorithm which is dependent on, and adapts to, the statistics of the image being encoded. Nonadaptive coding techniques are usually designed to compress the average or typical image (e.g., text) very efficiently, and consequently are less efficient for the atypical image.

As indicated by Figure 6.1.1, most of the technical activity to date has been in the area of nonadaptive coding, which, in turn, is divided between techniques that reduce redundancy in one dimension, and those that reduce redundancy in two dimensions. The most well known one-dimensional code is the modified Huffman code, which is the CCITT standard for Group 3 facsimile equipment.

There are two different types of codes that remove redundancy in two dimensions: those which encode several scan lines simultaneously, and those which encode each individual scan on a line-by-line basis.

INFORMATION PRESERVING

+ Nonadaptive
  • One-dimensional
  — Modified Huffman

  • Two-Dimensional
  — Multiple Line Coding
  — Line-by-Line Coding
    + Predictive
      • RLC Prediction Errors
      • RLC Prediction States
      • RLC Reordered Data
    + Nonpredictive
      • Modified Read
      • Modified Read II

+ Adaptive
  • Arithmetic Coding

APPROXIMATION

• Interpolation
• Feature Recognition
• Symbol Recognition

Figure 6.1.1 Data compression techniques for black-white graphics.

The two-line Weber code is an example of a multiple-line coding technique.[1] It is generally accepted that the more aggressive line-by-line techniques outperform the multiple-line techniques.

There are two subclasses of line-by-line coding techniques: those which first perform a predictive function, and those which do not. The predictive function is best explained by referring to Figure 6.1.2 which shows one possible rule for predicting the brightness of a pel based on the preceding pel in the scale line and three adjacent pels in the line above. In such a system, the original binary image is processed according to the prediction algorithm to form a new binary "image," where each pel in the new image is either a correct or an incorrect prediction. The next question is how to process the new image of correct and incorrect predictions for transmission. Three alternatives have been proposed in the literature. The first performs a conventional run length coding of the prediction image.[2] The second performs a run length coding process on 16 different prediction images,[3] each devoted to a separate prediction state in Figure 6.1.2. The third performs a reordering function prior to run length coding.[4] In general these predictive techniques do not perform as well as aggressive nonpredictive techniques.

The nonpredictive coding techniques directly process the imagery data on a line-by-line basis. The most well known example of this compression technique is the modified read code, which is the optional code in Group 3 facsimile. (The compression algorithm is described in detail in Chapter 2.) This coding concept is

State Number		Prediction	Probability of Correct Prediction	Probability of State	State Number	State Definition	Prediction	Probability of Correct Prediction	Probability of State
1	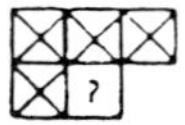	B	.96	.055	9	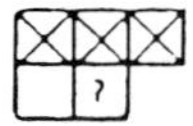	W	.63	.017
2	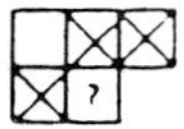	B	1.0	.0076	10	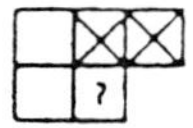	B	.71	.026
3	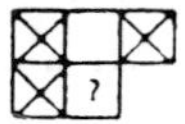	B	1.0	.0001	11	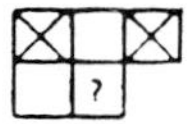	W	1.0	.0003
4	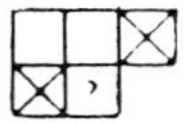	B	.99	.0033	12	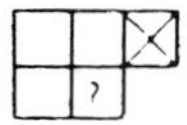	W	.85	.032

Figure 6.1.2 Four pel prediction algorithm.

State Number		Prediction	Probability of Correct Prediction	Probability of State	State Number	State Definition	Prediction	Probability of Correct Prediction	Probability of State
5	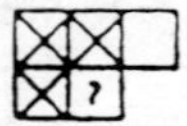	B	.83	.0291	13	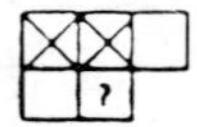	W	.84	.003
6	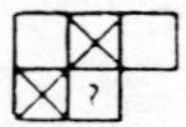	B	1.0	.0008	14	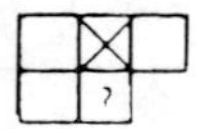	B	.86	.0015
7	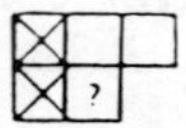	W	.74	.027	15	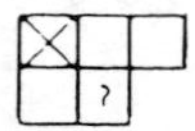	W	1.0	.008
8	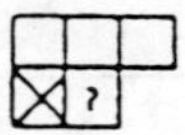	B	.55	.017	16	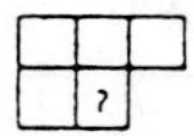	W	.99	.77

Figure 6.1.2 (continued).

based on the transmission of the location of black-white transitions, relative to a corresponding transition on the line above.

The basic MR code is designed for facsimile transmission over the PSTN and between fax machines which are mechanically limited in their ability to advance the paper rapidly when transmitting a redundant (e.g., all-white) portion of the page. For example, if a large portion of the page is white, the network is frequently required to transmit "fill" bits because the fax machine is slower than the compressed data. The modified modified read code is an adaptation of the basic MR code for transmission over data circuits with error control and between fax machines with image storage. Table 6.1.1 compares the characteristics of the two modified read codes.

An example of an adaptive information-preserving technique is arithmetic coding. This technique permits the more efficient compression of binary sequences as the statistics of the facsimile data change.[5,6]

Approximation Techniques

In the case of approximation techniques the output binary image is usually not a perfect replica of the input binary image. Approximations are made for the purpose of achieving a higher compression ratio than can be obtained from those techniques that exactly reproduce the input image.

In the case of interpolation techniques, particular pels in the input image are deleted from transmission. At the receiver, the brightness values of the missing pels are estimated from the adjacent pels that are transmitted. In the case of line interpolation, every second line is deleted from transmission, and the missing lines are replaced at the receiver by a line interpolation algorithm. (A brief description of such an algorithm is provided in Section 6.3.)

The ultimate compression technique, which reproduces an approximation of the input image, is a system that recognizes symbols such as alphanumeric text. A feature recognition system is very similar to symbol recognition; the primary difference is the complexity and size of the pattern recognized. In the case of feature recognition, the pattern recognized could be a portion of a symbol such as the crossbar of the t or the vertical bar in H. Figure 6.1.3 (a) illustrates a set of five symbols sampled into binary pels. Figure 6.1.3 (b) illustrates the same symbols after the binary pels have been processed by an algorithm that divides the symbol into quadrilateral features. A high degree of compression is achieved by transmitting merely the coordinates of the corners of the quadrilateral features.

The ultimate graphic compression technique is symbol recognition, whereby the large number of pels defining a symbol are transmitted by a short ASCII-like code. One basic concept describing this approach is *combined symbol matching* (CSM),[7] where the system adapts to the particular symbol font that is transmitted.

Table 6.1.1 Difference between Group 3 two-dimensional coding scheme and Group 4 basic coding scheme.

ITEM / APPARATUS	First line	K parameter	Line synchronization code word	End of page	Fill bits per line	Pad bits per page	Run length longer than 2623
G3	one-dimensional coding (MH)	K=2 (standard Res.) K=4 (optional higher Res.)	EOL+tag	RTC =6×(EOL + tag)	variable length string of "0"s	Not specified	Not specified
G4	two-dimensional coding (MR) Note : reference line for first coded line is set to an imaginary white line	K : not specified all lines including the first line of a page are coded two -dimensionally	Not required	EOFB =2×EOL	Not required	variable length string of "0"s	Using the MUC (s) of 2560

(a) Image definition by pels.

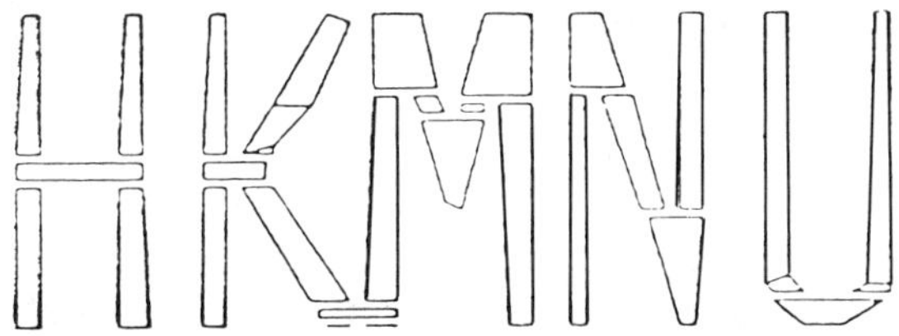

(b) Image definition by quadrilateral features.

Figure 6.1.3

This is done by first building up a library of characters at both the transmitter and receiver whenever a new symbol is encountered. When a subsequent symbol is found to match closely a symbol in the library, a short ASCII-like code is transmitted to the receiver. At the receiver, the corresponding bit pattern is retrieved from the library and printed. A basic requirement of symbol recognition systems is that the image must be divided into two parts, symbols and graphics. In the case of CSM, the graphic portions of the image are coded by using the MR algorithm.

6.2 INPUT SPATIAL FILTERING

Figure 6.2.1 is an enlarged view of typographical characters illustrating the ragged nature of the edge of the image. Notice that the raggedness varies from character to character, indicating the complex and variable relationship between the paper fiber and the printing process. In the conventional printing process, this raggedness is obviously not visible to the eye due to its small size. Figure 6.2.2 illustrates how the scanning and spatial sampling process inherent in digital facsimile systems creates a much higher level of raggedness than existed in the original image. This digital "noise" is not only harmful from an image quality point of

Figure 6.2.1 Enlarged view of typographical characters.

view, but it also reduces the compressibility of the source image. Therefore, spatial filtering of the scanned fax signal prior to compression would be advantageous to improve the output picture quality and to reduce transmission time.

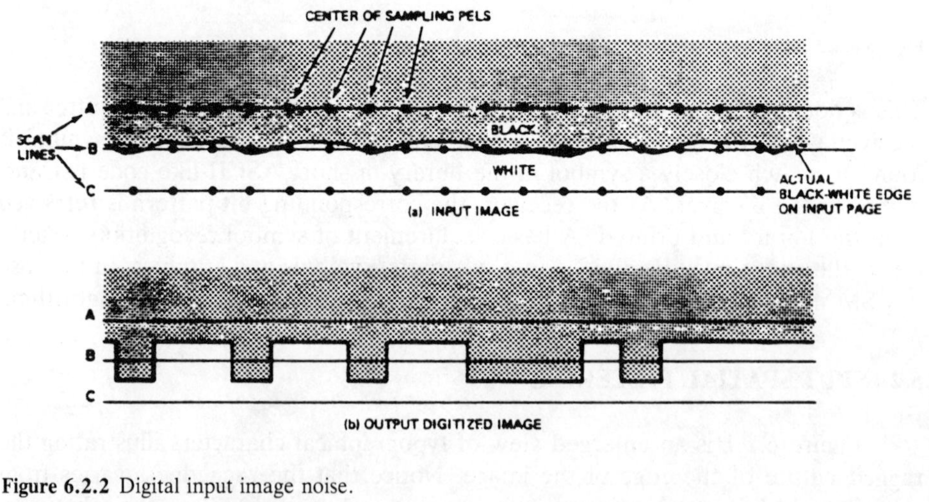

Figure 6.2.2 Digital input image noise.

The performance of a simple spatial filter was measured by means of computer simulation. The filter is illustrated in Figure 6.2.3 for a particular orientation and color combination. The three other orientations and four other color combinations are included in the simulation. Figures 6.2.4 and 6.2.5 illustrate the input and output pixel image for such a filter. The compression for the filtered image was improved between 5 and 10% relative to the unfiltered image.

Figure 6.2.3 Spatial filter.

6.3 INTERPOLATION

Source coding, a technique that reduces redundancy inherent in the input document, is the most commonly used image compression technique. The modified Huffman and modified read codes are examples of source coding techniques. A fundamental characteristic of source coding is that the compression ratio is highly dependent on the complexity of the input document. Complex documents require more transmission time than do documents containing little information. Interpolation techniques, however, achieve a fixed compression ratio independent of picture content. Consider the alternate line interpolation method, which achieves a fixed compression ratio of 2:1 by deleting alternate scan lines from transmission. At the receiver, the missing scan lines are interpolated from the transmitted lines.

Consider a typical digital facsimile system, which scans and prints with the same resolution in both the horizontal and vertical directions. For example, the high resolution option for Group 3 facsimile equipment employs a resolution of 7.7 lines/mm in both directions. Figure 6.3.1 illustrates the general concept of line interpolation where each square represents one pixel in the input and output images (e.g., 1/200 inch × 1/200 inch). As shown in the diagram, every second line of pixels is transmitted, and the lines not transmitted are interpolated at the receiver.

In Figure 6.3.1, the pixel labelled "x" is interpolated by using the transmitted information from the lines above and below. The interpolation scheme differs from source coding in two respects. First, the compression ratio is fixed at the 2:1, regardless of the redundancy of the input image. Second, the interpolation technique does not reproduce the input digital image exactly, and thus errors will occasionally be made.

The above discussion views the proposed system from a data compression perspective. We may also consider the interpolation proposal from an image enhancement viewpoint. Figure 6.3.2 illustrates the typical "stairstep" distortion in a conventional digital facsimile system when scanning a black-white edge, which is nearly parallel with the scanning track. This ragged edge distortion is highly visible to the naked eye and significantly reduces the acceptability of digital facsimile. We can enhance the appearance of the output image by artificially creating, or interpolating, twice as many scan lines on the output copy as were scanned and transmitted. If this were done, the size of the ragged stairstep distortion would be

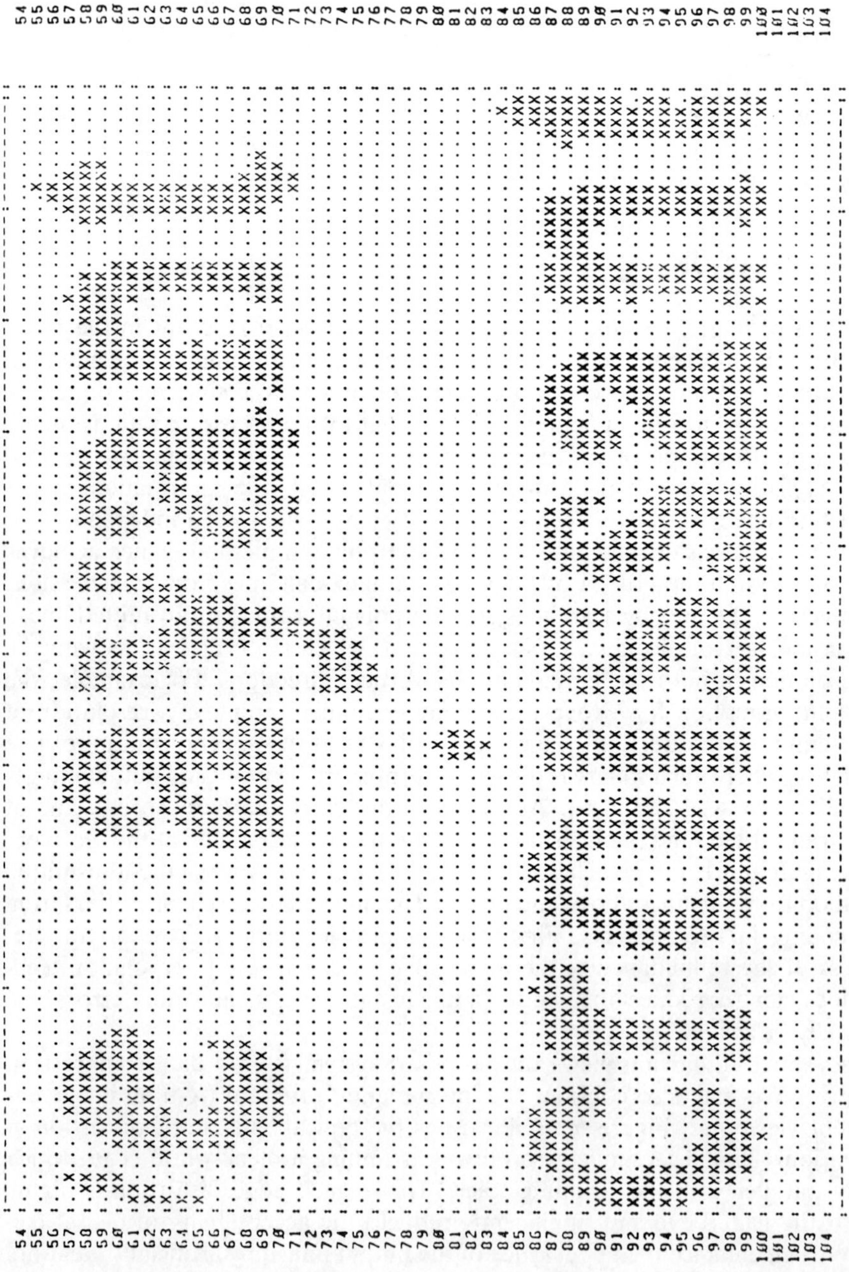

Figure 6.2.4 Unfiltered image.

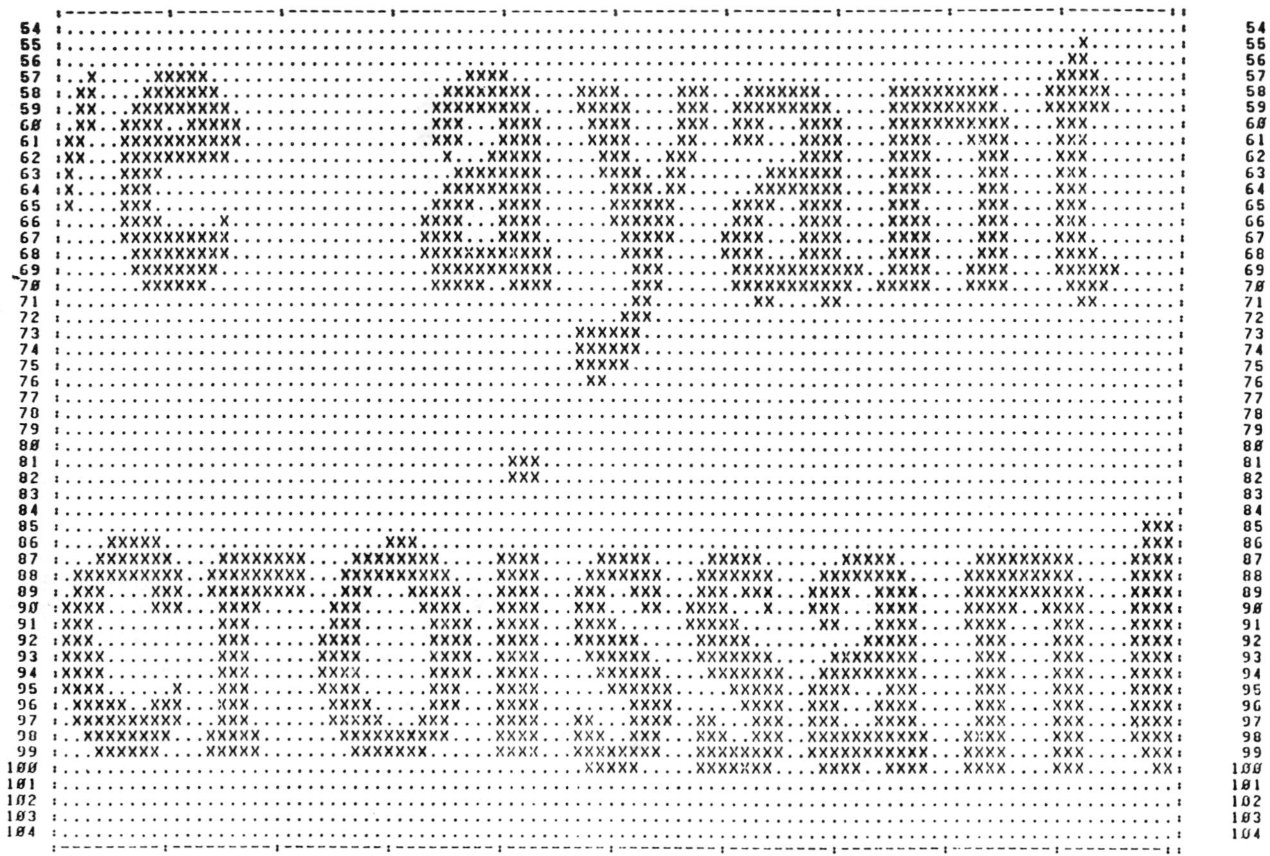

Figure 6.2.5 White and black single pel on edge.

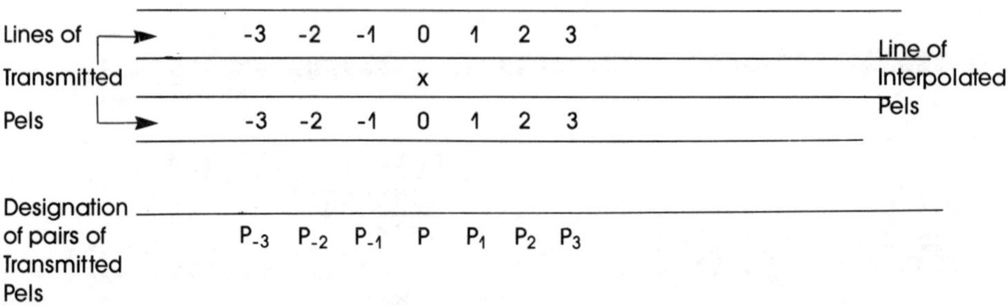

The pel to be interpolated is designated as x. The first step in the interpolation process is to consider the pair of transmitted pels (P_0) which are adjacent to x in the line above and below. If the P_0 pels are both black, interpolate x to be black. If the P_0 pels are both white, interpolate x to be white. If the two P_0 pels are of different color, adjacent pairs are examined to determine the color of the nearest pair where both pels are black or white. Pel x is interpolated to be the color of the nearest pair which is all black or all white. For example, if pairs P_{-2}, P_{-1}, P_0, and P_1 all have differing colored pels in each pair, but the pels in P_2 are both white, then x is interpolated to be white. If opposite colors are found equidistant from x, then x is interpolated to be white. For example, under the following conditions x is interpolated to be white:

P_{-2} — both black
P_{-1}, P_0, P_1 — pairs have different colored pels
P_2 — both white

Figure 6.3.1 Alternate line interpolation.

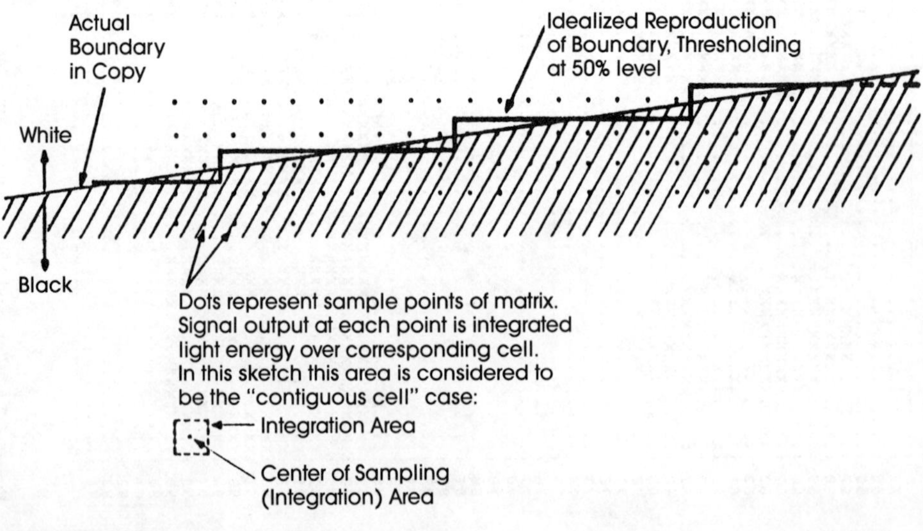

Figure 6.3.2 Staircase.

reduced to half, and the image would be more pleasing to the eye. The proposed interpolation algorithm would automatically position the stairstep transitions on the interpolated lines in the optimum position, halfway between the steps on the transmitted lines above and below.

6.4 PATTERN RECOGNITION

A large fraction of the pages transmitted via facsimile systems contain primarily textual information. Several investigators have proposed pattern recognition techniques to detect symbols, such as text in an OCR-like manner, and to transmit the location and identification of the patterns by using a short, efficient, ASCII-like code. Conventional OCR techniques are not feasible in the facsimile environment due to the large number of fonts that could be encountered and the uncontrolled nature of the fax signal. It is feasible, however, to consider an adaptive OCR-like system, which develops a library of characters as the page is scanned. The first work in this pattern recognition area was identified as combined symbol matching.[7] More recent work has been performed by AT&T, culminating in a technical contribution to the CCITT in 1983. The remainder of this section describes the AT&T technique and its performance.

A facsimile image is examined line by line. When a black pel is located, a *pattern isolator* surrounds connected pels to extract the entire pattern of which the pel is a member. The pattern is decomposed into $M \times M$ sub-blocks called *symbols*. In many cases, a symbol will be an isolated pattern that entirely fits within the $M \times M$ block. The incoming symbol is matched with already identified symbols stored in a library. If a match is detected, information is coded about the position of the symbol in the image and in the library. If no match is found, the incoming symbol is added to the library, which is assumed to be empty at the beginning of the coding and is gradually built up. In this case, an accurate description of the symbol and its position in the image is coded. A block diagram of the coding algorithm is presented in Figure 6.4.1.

The performance of the AT&T pattern recognition system was measured by means of computer simulation. Compression ratios were measured for three of the CCITT test documents at four different resolutions. The compression of the pattern recognition system is approximately three times that for modified modified read.

This higher compression is not achieved without a cost in picture quality. Figure 6.4.2 illustrates the pixel printout for the bottom of CCITT document number 1 as it would be printed for a lossless compression system, such as MR2. Figure 6.4.3 shows the comparable image produced by the pattern recognition system. Notice that the letters e and s have all been recognized to be the same character. It should be noted, however, that the characters on the input document are very small, much smaller than typed characters. It is likely that the misrecognition would not affect the readability or intelligibility of the document at all.

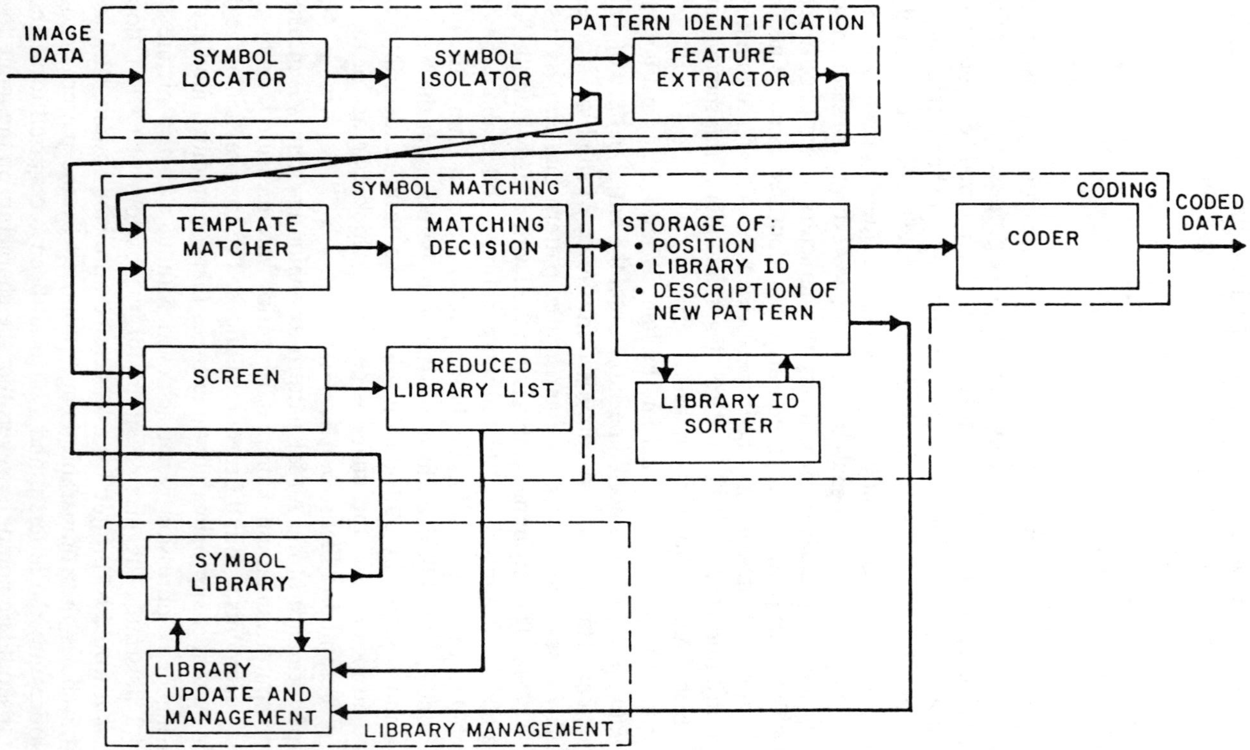

Figure 6.4.1 Block diagram of the AT&T pattern matching coder.

Figure 6.4.2 Original input image: CCITT Document #1 — 200 lines per inch, bit image printout.

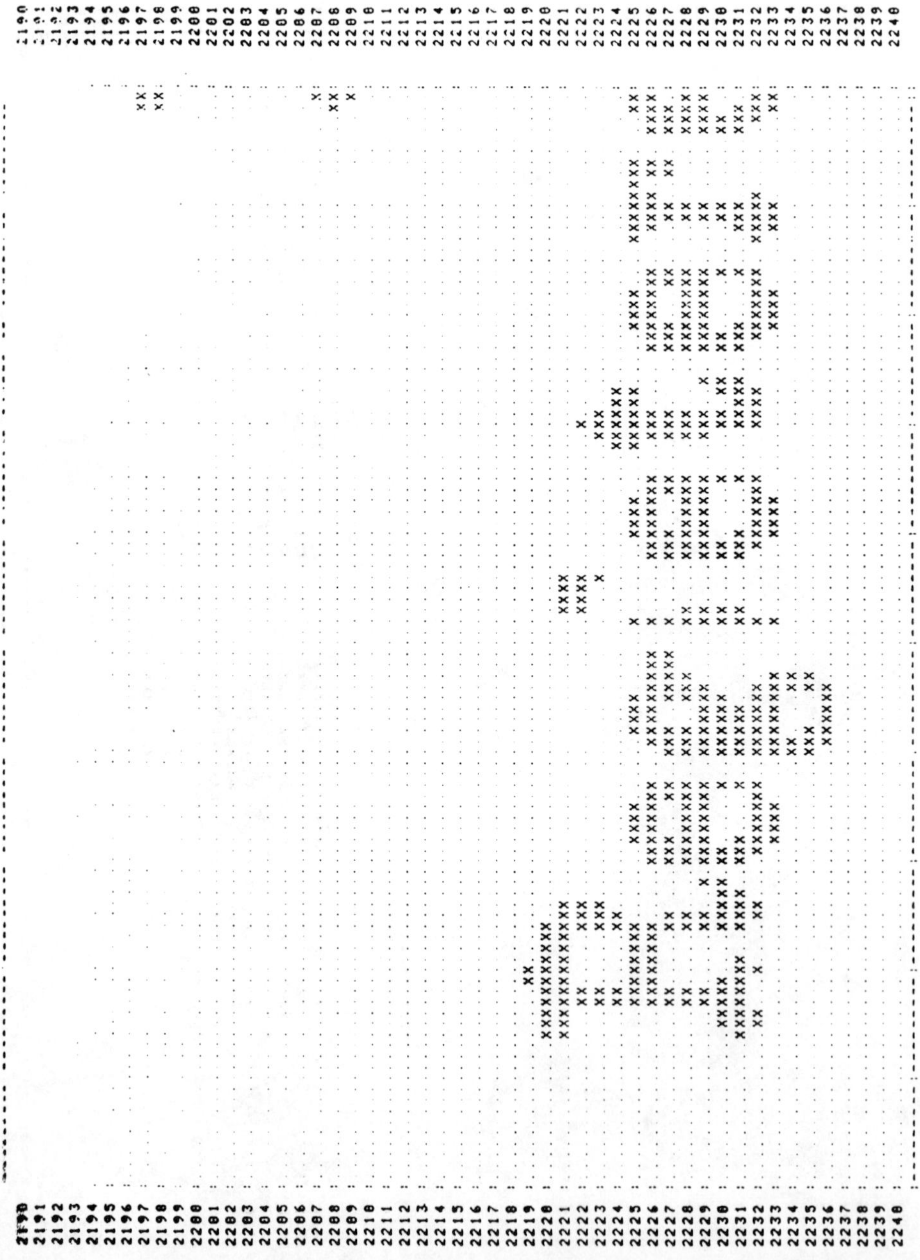

Figure 6.4.3 Output image: CCITT Document #1 — 200 lines per inch, bit image output, reject threshold — 4.

6.5 ADVANCED GROUP 4 CODING FOR BINARY IMAGES

The basic coding scheme for Group 4 facsimile is a two-dimensional line-by-line technique, which is merely an extension of the MR algorithms for Group 3 facsimile. There is a great deal of activity by the standards organizations toward improved coding of binary images relative to the existing standard. Most of the new work on compression addresses facsimile applications that differ radically from conventional message fax. In one case, the Open Document Architecture (ODA) will be used for narrowly defining the structure of mixed-mode documents. This would be used primarily where the documents originate in a computerized publishing environment rather than operating on scanned raster data. In a second scenario, the ISO and the CCITT have formed a joint committee called the Joint Bi-level Image Group (JBIG) to develop a new standard for black-white imagery. The primary application of the new algorithm will be the exchange of binary images over the ISDN with considerable emphasis on access by CRT rather than printers.

Figure 6.5.1 is an example of a layout structure of a letter as defined by the ODA. A coordinate system has been defined for a page based on basic measurement units (BMUs), where there are 1200 BMUs per inch and the point of origin is the upper left-hand corner. An image is defined by locating *blocks* and *frames* of image data on the page as illustrated in the figure. The ODA is extremely efficient for image compression because no data are transmitted for the blank parts of the page. A high level of compression is also achieved because text information is usually inserted by key stroke, permitting the definition of a character by the highly efficient ASCII code. The ODA also provides for the ability to define graphic imagery by means of vectors rather than by raster, yielding an additional degree of compression for graphics.

As indicated above, the JBIG committee is developing a standard for coding bi-level pictures for efficient retrieval and viewing on soft copy displays over 64-kb/s ISDN data links. This standard would be used for the rapid retrieval of documents from a database and the real-time interchange of pages from desk to desk. While conventional facsimile is transmitted in a line-by-line sequential manner, the JBIG algorithm is based on progressive transmission, where a low-resolution version of the entire page is first presented, followed by resolution enhancement over time. In this way, a user browsing through a document could quickly skip to the next page after having seen only a very low resolution version of a page.

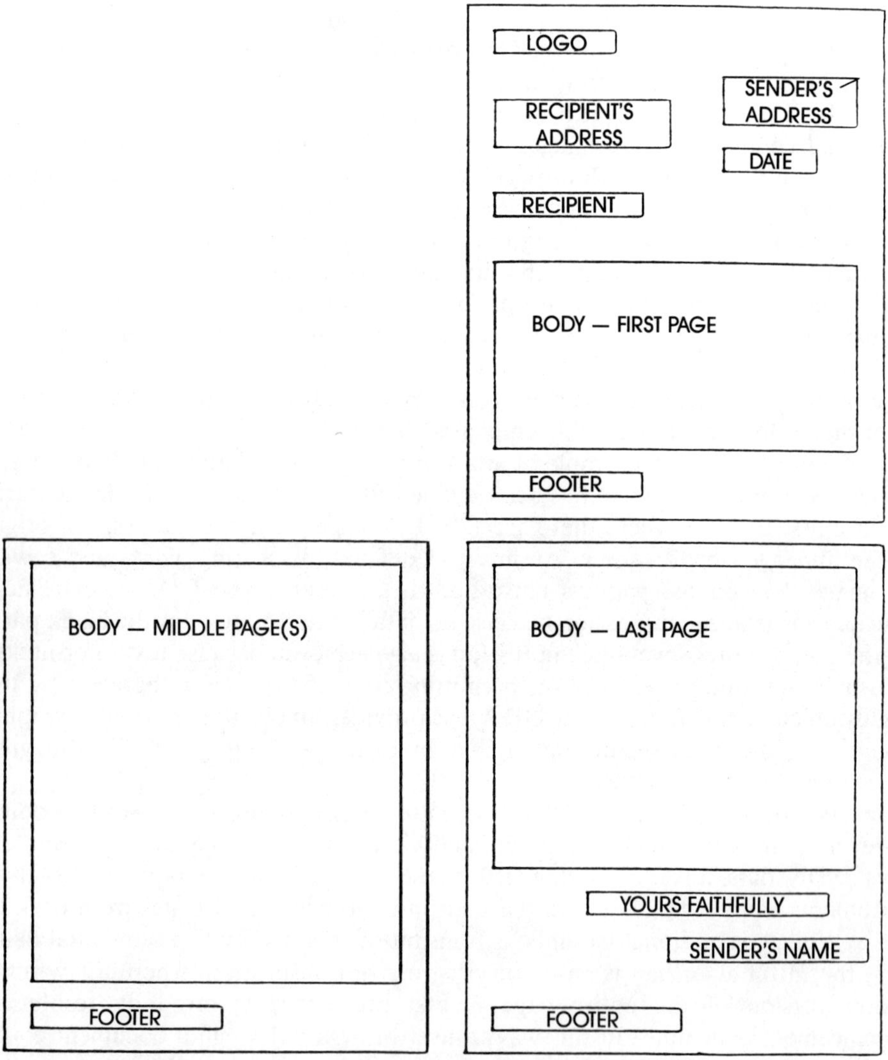

Figure 6.5.1 Example of a layout structure for a document called "letter."

REFERENCES

1. Hochman, D., and Weber D.R., "Dacom Facsimile Data Compression Techniques," *Proceedings International Conference Communications,* 1970, pp. 20.14–20.21.

2. Federal Republic of Germany, "Two-dimensional Coding Scheme," CCITT SGXIV Doc. No. 82, March 1979.
3. Xerox Corporation, "Proposal for an optional two-dimensional coding apparatus," CCITT SGXIV Doc. No. 84, April 1979.
4. Netravali, A.N. and Mounts, F.W., "Ordering Techniques for Facsimile Coding: A Review", *Proceedings IEEE,* Vol. 68, No. 7, July 1980, pp. 796–807.
5. Langdon, G.G., Jr. and Rissanen, J., "Compression of Black-White Images with Arithmetic Coding," *IEEE Transactions on Communications,* Vol. COM-29, No. 6, June 1981, pp. 858–867.
6. Langdon, G.G., Jr., "An Introduction to Arithmetic Coding," *IBM Res. Develop.* Vol. 28, No. 2, pp. 135–149 (March 1984).
7. Pratt, W.K., *et al.,* "Combined Symbol Matching Facsimile Data Compression System," *Proceedings IEEE,* Vol. 68, No. 7, July 1980, pp. 786–796.

Chapter 7
Coding Gray Scale and Color Pictures

7.1 OVERVIEW

The compression of binary black-white images was reviewed in Chapter 6. The purpose of this chapter is to discuss techniques for coding continuous-tone gray scale images and color pictures. Figure 7.1.1 is a hierarchical chart, which classifies the wide range of gray scale coding techniques that are available to the designer of facsimile systems.

Gray scale coding is usually accomplished in a two-step process. The input analog signal is first quantized, or converted to digital form, by encoding either individual pels or blocks of pels. The probability distribution of the encoder output levels is usually highly nonuniform, indicating that there is residual redundancy. This redundancy is reduced by assigning variable length codes to the coder output: short codes for frequently occurring levels and long codes for unlikely conditions. This class of coding is defined in Figure 7.1.1 as the statistical code assignment technique.

There are two types of quantization technique in which individual pels are encoded: pulse code modulation (PCM) and predictive coding. PCM is the oldest and most basic gray scale coding technique in which the image signal is sampled at the Nyquist rate, each sample is quantized to 2^m levels, and each level is represented by a binary word m bits in length. Because each pel is encoded independently of its neighbor, no redundancy is reduced. A pel is typically quantized with eight-bit precision to avoid the possibility of any visible distortion. For this reason, eight-bit PCM coding is commonly used as a reference to measure the performance of image compression techniques.

Predictive coding is a basic workhorse compression technique, which encodes a picture pel by pel. In this technique, the brightness value of each new pel is predicted on the basis of previously coded information that has been transmitted. The actual brightness value of the new pel is subtracted from the predicted value, and the differential error signal is quantized for transmission. A significant bit-rate reduction can be achieved by predictive coding for two reasons: first, the error

PEL CODING

- PCM
  - Fixed
  - Adaptive
- Predictive Coding
  - Quantization
    - Fixed
    - Adaptive
  - Prediction
    - One-Dimension
    - Two-Dimension

BLOCK CODING

- Transform Coding
  - Transform
    - Karhunen-Loeve
    - Discrete Cosine
    - Discrete Fourier
    - Walsh-Hadamard
  - Coefficient Coding
    - Fixed
    - Adaptive
- Vector Quantization

STATISTICAL CODE ASSIGNMENT

- Fixed (Variable Length Coding)
  - Huffman
  - Shannon-Fano
- Adaptive
  - Arithmetic Coding

OTHER

- Bit Plane Coding
- Stanag 5000
- Interpolation

Figure 7.1.1 Classification of gray scale coding techniques.

signal is statistically more redundant than the original pels; second, the eye is less sensitive to coarsely quantizing the error signal than the original PCM pels. There are a wide range of predictive coding systems based on different designs for the quantizer (fixed or adaptive) and the predictor (one-dimensional or two-dimensional prediction).

In the other major category of coding techniques, the image is encoded in blocks of pels rather than on a pel-by-pel basis. *Transform coding* and *vector quantization* are examples of techniques that encode blocks of pels. In the case of transform coding, a block of pels is transformed into another domain, called the *transform domain,* prior to quantization and transmission. Significant bit-rate reductions can be achieved in transform coding for two reasons: first, a large fraction of the coefficients can be deleted from transmission without affecting picture equality; second, the transform coefficients can be encoded with less accuracy than the original pixels without severe results for picture quality.

The conventional scalar quantization process is accomplished in two steps: first, the range of possible input values is divided into subranges or subsets; second, a single representative value is printed or displayed at the output when an input value falls within a particular subrange. With vector quantization, the same two operations occur, not in one-dimensional scalar space, but in an N-dimensional vector space. (This technique will be discussed further in Section 7.5.)

The CCITT and ISO have formed a committee to develop a standard gray scale coding technique for a wide range of applications such as Group 4 facsimile, freeze frame TV, and videotex. This committee, the Joint Picture Encoding Group (JPEG) illustrated in Figure 7.1.2, has rigorously compared several competing algorithms, such as the *discrete cosine transform* and predictive coding. The primary objective of the group is to develop a standard for a progressive coding algorithm for the rapid distribution of pictures over the 64-kb/s ISDN channel.

7.2 PULSE CODE MODULATION (PCM)

In PCM coding, the signal is sampled at the Nyquist rate, each sample is quantized to 2^m levels, and each level is represented by a binary word m bits in length. At the decoder, these binary words are converted to a series of amplitude levels which is low-pass filtered. Because each sample is encoded independently of its neighbor, no redundancy is reduced, and PCM is used as a reference to measure the performance of bit-rate reduction techniques. The PCM coding precision typically used for the reference is six to eight bits per pel because this precision ensures that there is no visible degradation in the picture. Figures 7.2.1, 7.2.2, and 7.2.3 illustrate the type of contouring distortion which occurs when the coding precision is limited, being 2-bit, 3-bit, and 4-bit precision, respectively.

The eye is logarithmic in its sensitivity to brightness distortion; that is, it is equally sensitive to equal percentage changes in brightness. For this reason, the PCM quantization is typically nonlinear with the size of the quantizer steps increasing with brightness. This nonlinear companding improves the picture quality for a given number of bits per pel.

Figure 7.2.3 illustrates the picture quality for conventional 4-bit PCM coding. A contouring type of distortion is visible and is somewhat disturbing because the eye is particulary sensitive to low frequency noise. The visibility of the distortion can be reduced by adding a low-level, pseudorandom, high-frequency noise to the signal prior to encoding, and subtracting the same noise at the receiver after

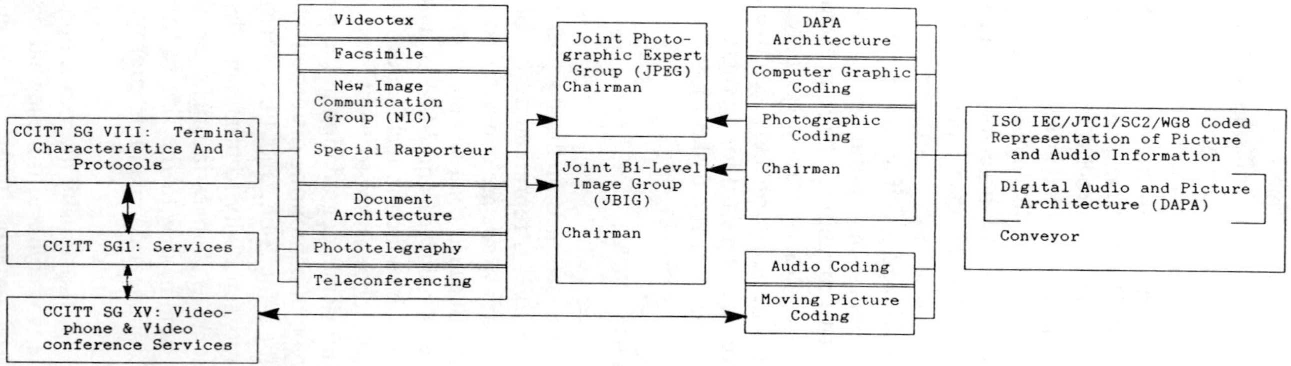

Figure 7.1.2 Organization of photographic standardization.

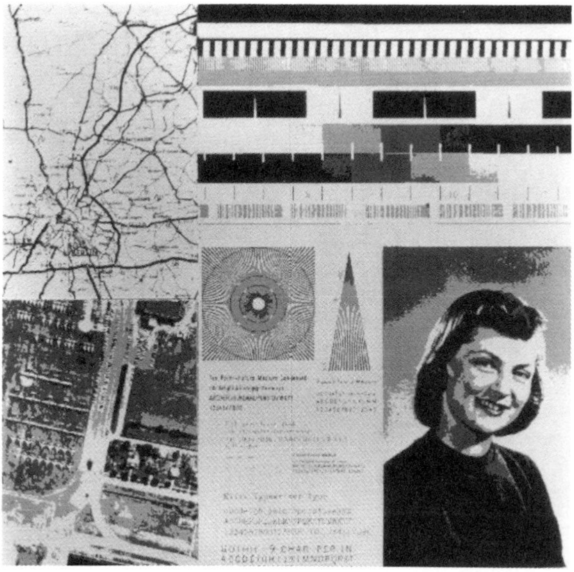

Figure 7.2.1 Two-bit PCM encoding.

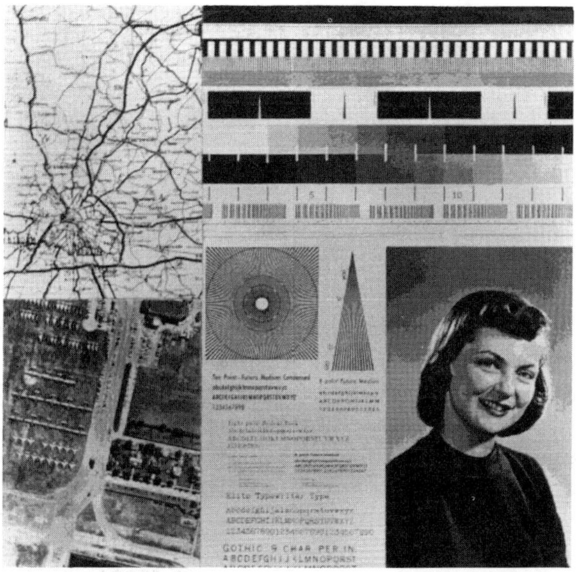

Figure 7.2.2 Three-bit PCM encoding.

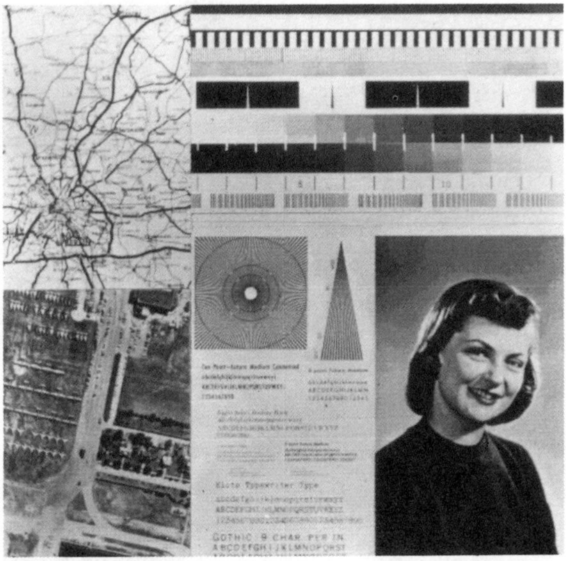

Figure 7.2.3 Four-bit PCM encoding.

decoding. The distortion associated with this technique, known as dither coding, is concentrated in the higher frequency region and is more acceptable to the eye. Figure 7.2.4 illustrates the quality of a 4-bit PCM dither coded picture.

7.3 PREDICTIVE CODING

PCM transmits each pel as an independent sample without taking advantage of the high degree of pel-to-pel correlation existing in most pictures. Predictive coding is a basic bit-rate reduction technique that does reduce this pel-to-pel redundancy. Figure 7.3.1 is a block diagram illustrating the basic predictive coding process. A predictor predicts the brightness value of each new pel based solely on the pels previously quantized and transmitted. The predicted brightness is subtracted from the actual value of the new pel, resulting in a bipolar prediction error signal. This error signal is quantized and transmitted. This quantization process can vary over a wide range of complexities. The most common technique is conventional pel-by-pel scalar quantization. We can, however, also employ transform coding or vector quantization. The quantization may be fixed, or it can adapt to the data. The quantizer can also vary over a wide range of accuracies. If one-bit quantization is employed, the system becomes the well-known *delta modulation* technique. If the predictive quantizer employs multiple bits per pel, the technique

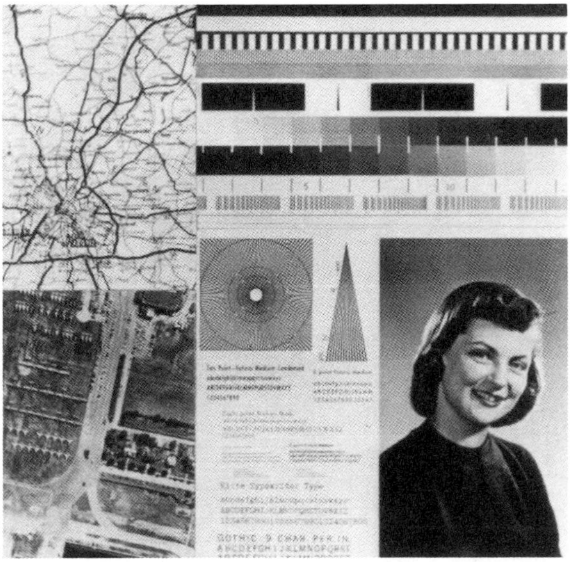

Figure 7.2.4 Four-bit PCM-PRN encoding.

is commonly defined as *differential PCM* (DPCM). At the receiver, the inverse of the quantization process is performed, and the decoded error signal is added to the predicted value to form the output signal for viewing. The output signal is fed to the predictor to be used for prediction of the next pel. Referring back to the predictive encoder, the reader will note that the transmitted signal is decoded at the transmitter using exactly the same decoding process which is used at the receiver. The predictive encoder can be viewed as a servo loop which continually forces the decoded output signal to be as close as possible to the input signal.

Figure 7.3.2 illustrates the transfer function of a typical three-bit DPCM predictive coder. The quantizer is usually nonlinear because the eye is very sensitive to small changes in low detail portions of a picture (small prediction error), but the eye is insensitive to coarse quantization of high-contrast edges (large predictive error). The design of this quantizer is a compromise between conflicting objectives. It is desirable that the quantizer precision be fine, particularly for small error signals, to keep the background granular noise in the output picture at an acceptably low level. On the other hand, the quantizer steps must be large enough, particularly the largest increment, so the output can respond reasonably well to high-contrast changes in the input picture. If the largest increment is too small, slope overload occurs resulting in picture blurring. Figure 7.3.3 is an illustration of the prediction error signal for a delta modulation system which exhibits severe slope overload.

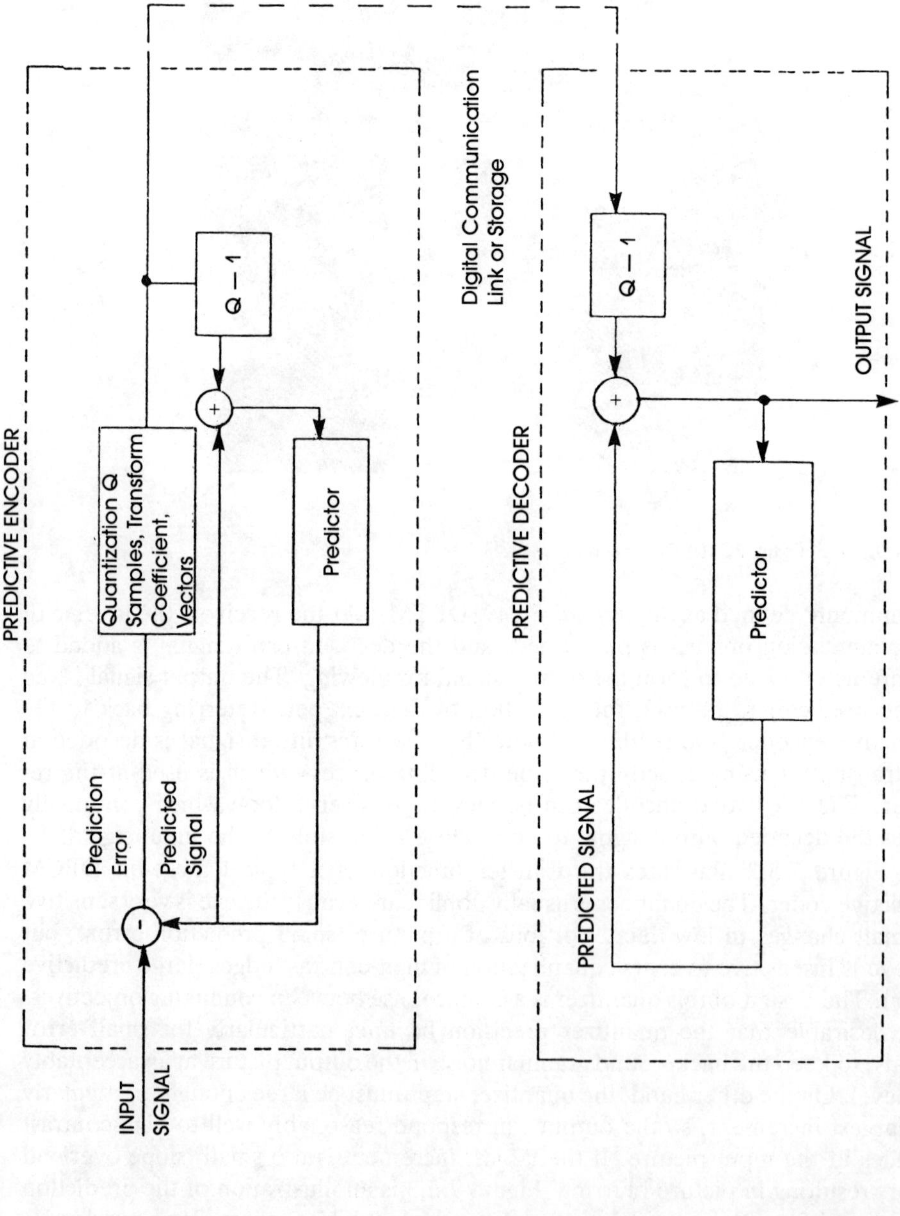

Figure 7.3.1 Block diagram of a generic predictive coding system.

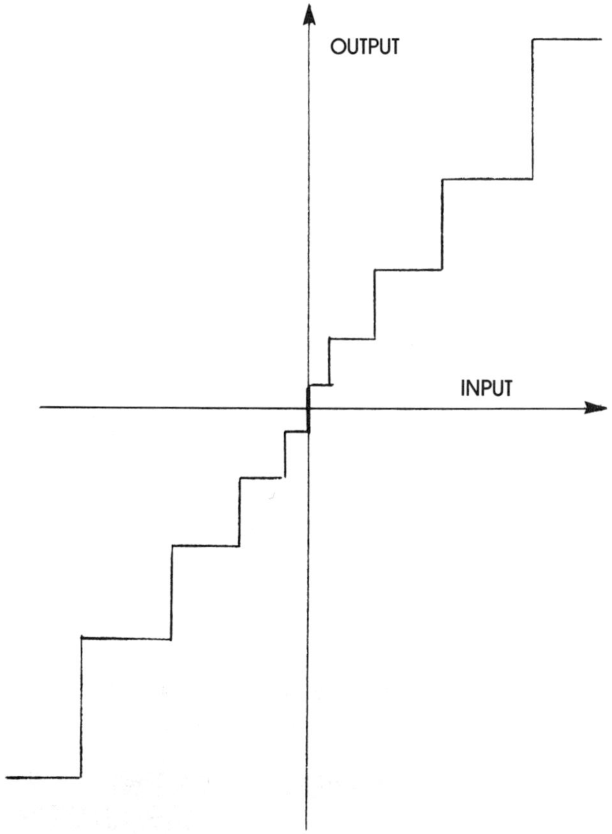

Figure 7.3.2 Quantizer for predictive encoder.

Figure 7.3.4 is a picture coded by using DPCM with two-bit/pel precision. Note that high-contrast vertical edges are slightly blurred due to the slope overload characteristic. Figure 7.3.5 also shows two-bit DPCM, but in this case two-dimensional rather than one-dimensional prediction is employed. Note that the vertical edges are sharper due to the more sophisticated predictor.

One fundamental characteristic of bit-rate reduction techniques such as predictive coding is that as redundancy is reduced the compressed signal is more sensitive to errors in the transmission channel. To illustrate this point, Figure 7.3.6 shows the output picture for five-bit PCM transmission if the bit-error rate is 10^{-2}. Figure 7.3.7 shows the output picture quality for a delta modulation system if the bit error rate is 10^{-3}. Clearly, the delta modulation scheme is much more sensitive to errors.

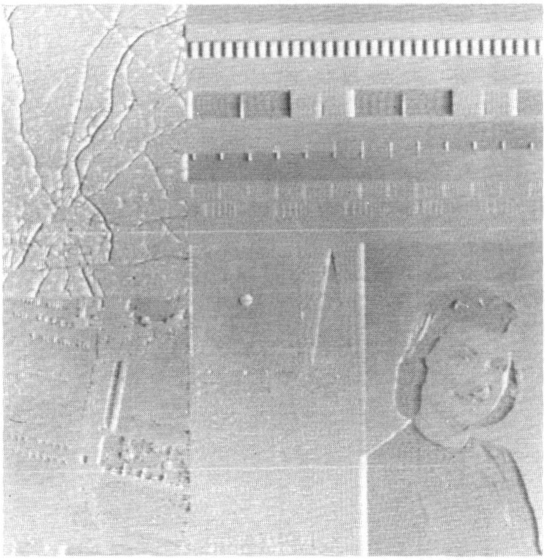

Figure 7.3.3 Severe slope overload.

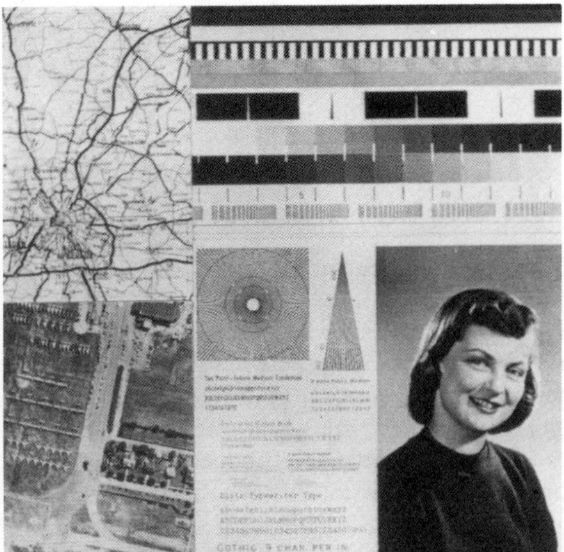

Figure 7.3.4 Two-bit delta modulation encoding.

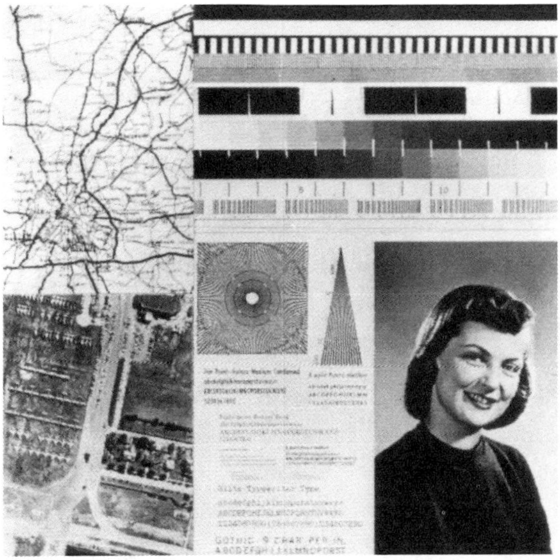

Figure 7.3.5 Two-bit H-V delta modulation encoding.

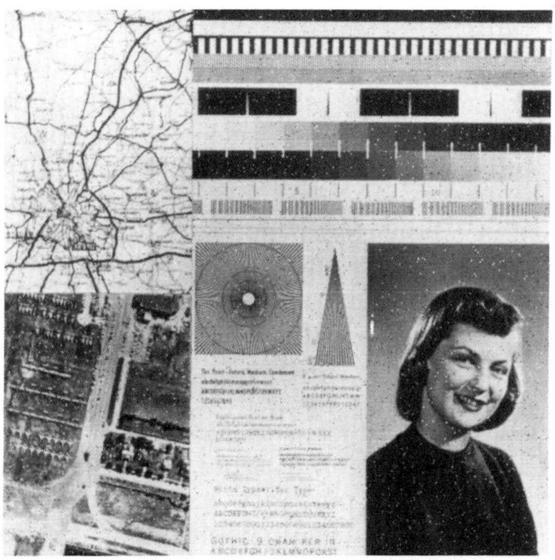

Figure 7.3.6 Five-bit PCM transmission. (Error rate = 10^{-2}.)

Figure 7.3.7 One-bit HIDM encoding with data link noise. (Error rate = 10^{-3}.)

7.4 TRANSFORM CODING

Transform coding algorithms, generally speaking, operate as two-step processes. In the first step, a linear transformation of the original signal (separated into sub-blocks of $N \times N$ pels each) is performed, in which signal space is mapped onto transform space. In the second step, the transformed signal is compressed by encoding each sub-block using quantization. The reconstruction involves performing an inverse transformation of each decoded and transformed sub-block. The function of the transformation is to make the transformed samples more independent than the original samples, so that the subsequent quantization may be done more efficiently.

The transformation operation itself does not provide compression; rather, it is a remapping of the signal in another domain where compression can be achieved more effectively. Compression can be achieved for two reasons. First, not all of the transform domain coefficients need to be transmitted in order to achieve acceptable picture quantity. Second, the coefficients that are transmitted can be encoded with reduced precision without seriously affecting image quality.

Transformation Techniques

Transforms that have proved useful include the Karhunen-Loeve, discrete Fourier, discrete cosine, and Walsh-Hadamard transforms. The Karhunen-Loeve transform (KLT) is considered to be optimal, and therefore many other transforms have been compared to it in terms of performance. The KLT, however, has certain characteristics that make it less than ideal for image processing. There is the necessity to estimate the covariance matrix before processing in both row and column operations. Also, the actual eigenvector determination must be carried out to generate the basis matrix. These drawbacks would not be significant if the efficiency of the KLT were much greater than that of other transforms. For data having high interelement correlation, however, the performance of other transforms (such as the discrete cosine) is virtually indistinguishable from that of the KLT, and thus usually does not warrant its added complexity.

The discrete Fourier transform (DFT) is one of the few complex transforms used in data coding schemes. There are disadvantages in using a complex transform for data coding, the most obvious being the storage and manipulation of complex numbers. As in the case of the KLT, this complexity would not be a factor if the performance of the DFT were significantly greater than that of other transforms. Other transforms which are less complex, however, perform better than the DFT.

The discrete cosine transform (DCT) is one of an extensive family of sinusoidal transforms. In discrete form, the basis vectors consist of sampled values of sinusoidal or cosinusoidal functions that, unlike those of the DFT, are real number quantities. The DCT has been singled out for special attention by workers in the image processing field, principally because, for conventional image data having reasonably high interelement correlation, the DCT's performance is virtually indistinguishable from that of other transforms that are much more complex to implement.

The three transforms mentioned previously have basis functions which are either cosinusoidal (i.e., the Fourier and discrete cosine), or are good approximations of a sinusoidal function (such as the Karhunen-Loeve transform). The Walsh-Hadamard transform is an approximation of a rectangular orthonomal function. The actual transform consists of a matrix of $+1$ and -1 values, which eliminates multiplication from the transform process. The elimination of multiplication is a significant property because the aforementioned transforms require real or complex multiplication. The Walsh-Hadamard transform, however, does not provide the excellent performance of the discrete cosine transform.

The discrete cosine transform is universally accepted as the preferred transform for image coding, and so it is useful to provide more detail on its implementation. The execution of the DCT algorithm requires the division of an image into a series of $(N \times N)$ sub-blocks of pixels. Each sub-block is transformed by a two-dimensional $(N \times N)$ DCT process as follows:

$$[T] = [C] \cdot [D] \cdot [C]^T$$

where $[T]$ is the transformed sub-block, $[C]$ is the DCT basis matrix, and $[D]$ is the input data sub-block ($[C]^T$ is the transpose of the DCT basis matrix). The DCT basis matrix coefficients were determined from the following relation:

$$C_{ij} = C_0 \cdot \sqrt{(2/N)} \cdot \{\cos[i \cdot (j + 0.5) \cdot (\pi/N)]\}$$

where $C_0 = 1/\sqrt{2}$ for $i = 0$, $C_0 = 1$, otherwise, and $i = j = 0$ to $N - 1$. Figure 7.4.1 illustrates the basis functions for a 16×16 DCT. This transformation converts each $(N \times N)$ sub-block of pixels into an $(N \times N)$ matrix of transform coefficients, which consists of one dc coefficient and $(N \times N - 1)$ ac coefficients. The sum of the squares of all of the ac coefficients in a given transform matrix is known as the ac energy of that transform matrix.

Coding of Transform Coefficients

As explained in the previous section, the DCT is usually used when pictures are transmitted by transform techniques. This transformation merely creates a set of coefficients equal in number to the original set of pels. At this point, no compression has been accomplished, except that the original set of pels with uniformly high redundancy have been decorrelated, and the information has been compacted in the lower spatial frequency coefficients. The purpose of this section is to address the second part of the two-step process: how to encode the transform coefficients for transmission.

Transform coefficients may be coded by either nonadaptive or adaptive techniques. Nonadaptive techniques are commonly used to encode TV pictures because the number of coded bits per image cannot vary over a wide range because the frame rate and the transmission rate are both fixed. A buffer permits the coded bits per picture to vary slightly, but the compression cannot vary over a wide range. In the case of facsimile systems, the transmission time is usually totally unconstrained; therefore, it is usually advantageous to employ adaptive coding techniques that are highly sensitive to the complexity of the pictures being transmitted. Complex pictures can take a much longer transmission time than simple pictures.

The first step in the coding process is to determine which coefficients are to be transmitted and which deleted. Figure 7.4.2 illustrates the set of 256 transform coefficients corresponding to a 16×16 block of pels to be coded. Coefficient 1 is the dc coefficient which is a measure of the average brightness of the block.

Figure 7.4.1 Two-dimensional cosine transform basis function.

Coefficients 2 through 16 measure spatial frequency content in the horizontal direction. Coefficients 17, 33, 49, *et cetera,* measure in the vertical direction, and all others are various combinations thereof. In general, most of the energy is contained in the low frequency coefficients with relatively little signal strength in the high frequency coefficients, as shown in Figure 7.4.3. In *adaptive zonal coding* the coefficients are transmitted in the zigzag sequence shown in Figure 7.4.2. A threshold is established, as shown in Figure 7.4.3, and a decision is made to truncate the zigzag sequence at the point where the coefficient amplitude falls below the threshold. An end-of-block code word is transmitted, indicating where the coefficients are truncated. The final step in the coding process is to quantize and to assign codes to those coefficients selected for transmission. This can be done by using codes having either fixed or variable length. In general, variable length codes are used to achieve the maximum compression.

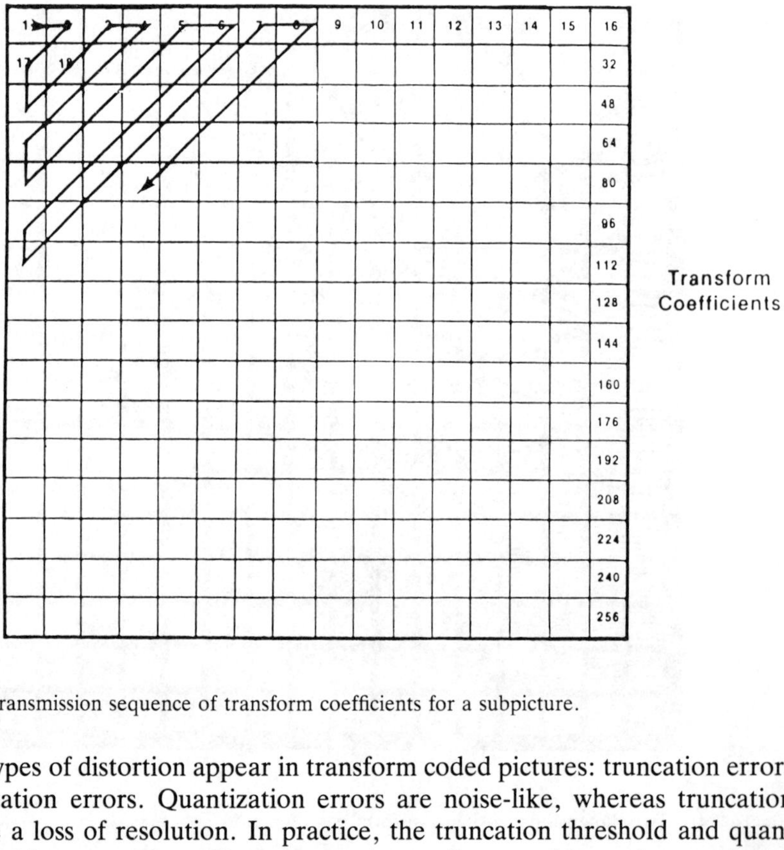

Figure 7.4.2 Transmission sequence of transform coefficients for a subpicture.

Two types of distortion appear in transform coded pictures: truncation errors and quantization errors. Quantization errors are noise-like, whereas truncation errors cause a loss of resolution. In practice, the truncation threshold and quantization precision must be adjusted experimentally to achieve the maximum compression and acceptable picture quality. In general, transform coding is preferable to predictive coding for compression to bit rates below one or two bits per pixel for single pictures. In applications where cost and complexity are important, however, the choice between these two algorithms may be less clear.

7.5 VECTOR QUANTIZATION

Vector quantization begins by dividing an image to be transmitted into rectangular blocks of pixels, with all blocks having the same dimensions. The transmitter compares each block with a large library of typical blocks, called a *codebook,* and selects the library block that best approximates the block to be transmitted. The transmitter then encodes and transmits the index to the selected library block. The receiver, equipped with a copy of the codebook, decodes the index, retrieves the selected library block, and inserts it into the output image.

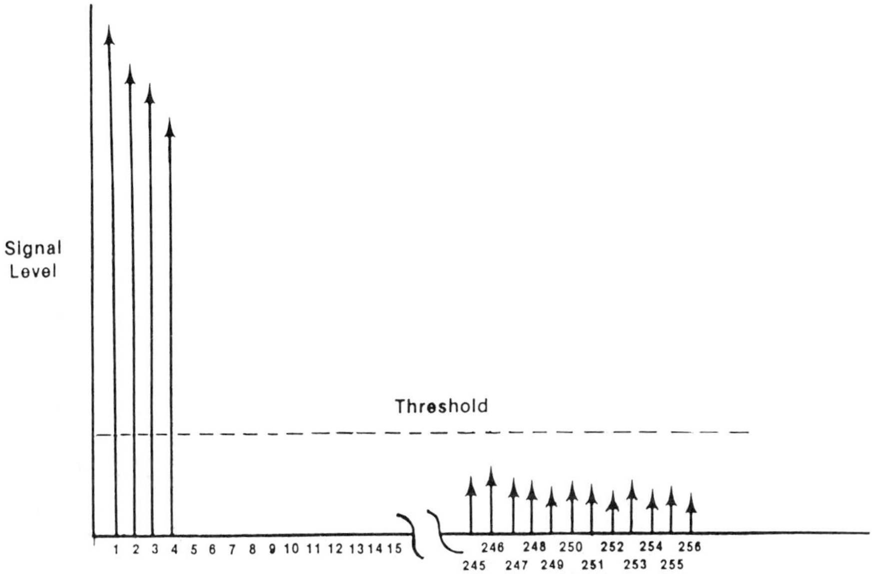

Figure 7.4.3 Transform coefficients.

This process is called vector quantization because, both theoretically and computationally, each block is treated as a vector. The vector representation of a block can be thought of as laying out all the gray scale values of the block pixels in a single string, that of the upper left pixel first, the lower right last. Such a string of numbers constitutes a vector in k-dimensional space, where k is the number of pixels in the block. When the block is treated in this manner, the entire body of mathematical knowledge of vector analysis and multidimensional analytical geometry can be applied to the vector quantization problem. In the balance of this discussion, *block* will refer to a rectangular array of pixels in an image, and *vector* will refer to the representation of these pixels as a string of numbers.

In all of the variations of vector quantization there is a trade-off between image quality and data compression. In the theoretical limit of zero distortion, the codebook would contain vectors representing all possible blocks and an exact match would always be found. Distortionless transmission would, however, entail an enormous codebook and little data compression, even with optimal coding. At the other extreme, a codebook containing few vectors (i.e., representative blocks) would yield large compression ratios, but poor image quality. The objective of any vector quantization system design is, therefore, to achieve the best compromise among codebook size, data compression, and received image quality.

A review of published papers revealed many variations on the vector quantization theme. Gersho[1] presents a mathematical treatment of the problem. The codebook is, in effect, the vector quantizer in that it quantizes the multidimensional vector space into a finite set of representative vectors. Gersho then explores the partitioning problem, and concludes that the only practical way to design the quantizer (i.e., select the vectors to be included in the codebook) is to take advantage of vector clustering.

The basic vector clustering algorithm was thoroughly developed by Linde, Buzo, and Gray.[2] This algorithm, known as the LBG algorithm for its authors, takes advantage of the fact that the vector representations of image blocks tend to cluster in the vector space. A codebook containing vectors representing the cluster centroids offers the best compromise between codebook size and received image quality. The method consists of using a long sequence of "training" vectors to design the codebook. Gray and Linde[3] explain and compare several variations on the LBG codebook generation method, and, hence, compare the resulting performances of Gauss-Markov sources. In particular, the authors show that tree-assisted searches allow the use of codebooks much larger than those practical with exhaustive searches at the expense of a suboptimal codebook. The performance is only slightly degraded with respect to the exhaustive-search approach. Huang and Woods[4] discuss predictive vector quantization, which consists of a combination of predictive filtering and vector quantization. The purpose of the predictive filtering is to remove redundancy before vector quantizing the residue. A vector quantization method that offers great promise of good compression and low distortion is described in Japanese Annex 4.[5] This method combines DPCM and vector quantization. Other references include Helden and Boekee[6] and Gersho and Ramamarthi.[7]

Vector quantization in all its forms requires a large codebook of vectors from which one is selected for each block to be transmitted. Two very important issues are therefore: (1) *codebook search* and (2) *codebook generation.* There are two basic search methods: *exhaustive* and *tree-assisted.* The exhaustive method is guaranteed to select the codebook vector that best matches the input vector. This method is practical, however, only for very small block sizes because the search is much faster. A binary tree search begins with a choice between two codebook vectors that act as "keys" to the next search level. The selection of one of these keys leads to another two-way choice, which leads to a better approximation of the input vector, which leads to yet another two-way choice, and so on. This method, though much faster than the exhaustive search, may fail to find the best match because, once a two-way choice has been made in a given tree level, the search may be directed to a subtree that does not contain the best match. The general m-way tree search, in which an m-way choice is made at each decision level, gives better performance as the value of m increases, at the expense of longer search time. The exhaustive search is the limiting case of one M-way decision, where M is the total number of codebook vectors.

The codebook generation objective is a codebook that gives low image distortion while minimizing the codebook size, which is important, not only to minimize memory and search time, but also to achieve high compression ratios.

All codebook generation methods reported in the literature are variations on the LBG method. In principle, if we knew the statistics of all images to be transmitted, we could generate a codebook analytically. The most commonly used method, however, consists of using a large number of training vectors, each one representing a "typical" image block.

The following is a summary of the LBG codebook generation method. Assume, for the moment, a partially optimized codebook. Each training vector belongs to a codebook vector in that the training vector matches the codebook vector at least as well as it matches any other. (Ties are broken in various ways, depending on the specific method used.)

The codebook is updated to make each codebook vector the centroid of the set of training vectors that belong to it, thus minimizing the average distortion with respect to a set of training vectors. The update may, however, cause some of the training vectors that belonged to a given codebook vector before the update to belong to a different codebook vector afterward. Another iteration is therefore performed to compute new centroids, and the codebook is updated again. This process is repeated until there is no further improvement, or the improvement is less than some specified value.

This iterative method of codebook improvement leads to a *local* minimum of average distortion. A *slight* perturbation of the codebook vectors gives greater distortion. This method leaves the possibility that some large change to the codebook may give even less distortion; hence, the local minimum is not necessarily the global minimum (best possible codebook for the training vector set).

Codebook generation begins with one codebook vector that is the centroid of all the training vectors. This vector is then split into two vectors that are very close to each other. The splitting objective is to make the number of training vectors belonging to the two codebook vectors approximately equal. The codebook is then optimized, as described above. The two (now optimized) vectors are then split into four, and optimization is repeated. The process is continued until a codebook of the required size is achieved.

The typical block size employed in vector quantization systems is 4×4 pels, and the bit-rate reduction which can be achieved is comparable to transform coding. The vector quantization technology, particularly in the area of codebook generation and search, is not fully mature, and consequently few operational systems have been implemented. Nevertheless, vector quantization will play a significant role in gray scale coding in the future.

7.6 BIT PLANE CODING

Most of the picture coding techniques described in previous sections are inexact in that they do not usually transmit an exact replica of the original PCM picture. *Bit plane coding* (BPC) is a lossless coding technique which exactly reproduces the input image. BPC requires the storage of at least one complete scan line at the transmitter prior to encoding and at the receiver after decoding. Consider the case where a four-bit PCM image is to be transmitted. In BPC, the four bits for all of the pels in a scan line are not transmitted pel by pel, but sequentially in accordance with the coding precision. First, all of the most significant bits of all the pels in the line are transmitted. This is defined to be the most significant bit "plane." Then, the second most significant bit plane is transmitted, and so on until all four bit planes are transmitted. Bit-rate reduction is achieved because each plane is encoded for transmission by using a binary image compression technique (as presented in Chapter 6). At the receiver, all planes are reassembled in the normal multibit-per-pel word structure such that the image can be printed pel by pel. Figures 7.6.1, 7.6.2, 7.6.3, and 7.6.4 are pictures showing the bit planes for a four-bit PCM picture. Notice how the less significant bit planes are more noisy and difficult to compress than the most significant bit planes. This particular image was transmitted with an average of 0.6 bits/pel by using the modified read code to compress each plane. This yields a compression of 6.7:1. Compression can be improved further if the four bit planes are defined as a Gray code rather than using the conventional binary code.

The NATO countries have adopted a standard for coding gray scale images known as Stanag 5000. Pels are transmitted by using the two-line wobble pattern illustrated in Figure 7.6.5 (US Patent 3,849,592; Bernard M. Rosenbeck). The coding technique is the four-bit BPC concept described above. Each pel is defined by using the Gray code in Figure 7.6.6 rather than the conventional binary code. Compression is increased further by adaptively reducing the resolution in particular portions of a bit plane. Each 2 × 2 matrix of pels within the wobble scan (see Figure 7.6.7) is examined to determine whether there is high detail or low detail. If there is little detail, the block of four pels is transmitted as a single pel. The criteria for use of the low resolution mode are provided below.

- *Bit plane* 1 (most significant) — *Low resolution* never used;
- *Bit plane* 2 — *Low resolution* used if transition threshold is not exceeded;
- *Bit plane* 3 — *Low resolution* used if transition threshold not exceeded *or* if bit plane 2 uses low resolution;
- *Bit plane* 4 — *Low resolution* always used.

7.7 COLOR FACSIMILE

There are very few color facsimile units available in the marketplace today. There are several reasons for this lack of activity including the complexity and cost

Figure 7.6.1 One-bit PCM.

Figure 7.6.2 Second most significant bit.

Figure 7.6.3 Third most significant bit.

Figure 7.6.4 Least significant bit.

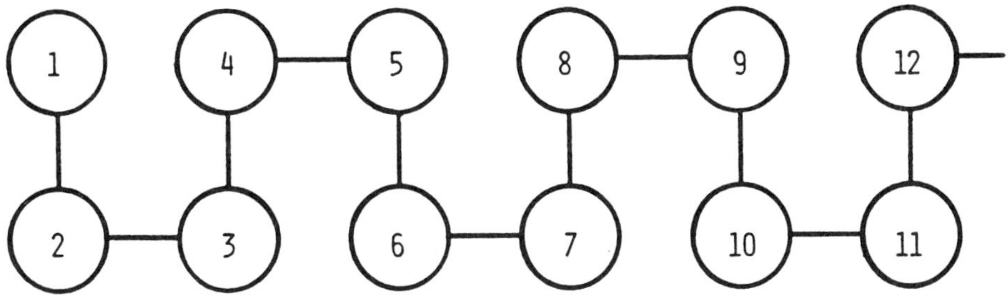

Figure 7.6.5 Two-line wobble scan.

Intensity Level	Normal Binary				4-bit Gray Code			
0	0	0	0	0	0	0	0	1
1	0	0	0	1	0	1	0	1
2	0	0	1	0	0	1	0	0
3	0	0	1	1	0	1	1	0
4	0	1	0	0	0	0	1	0
5	0	1	0	1	0	0	0	0
6	0	1	1	0	1	0	0	0
7	0	1	1	1	1	0	1	0
8	1	0	0	0	1	0	1	1
9	1	0	0	1	0	0	1	1
10	1	0	1	0	0	1	1	1
11	1	0	1	1	1	1	1	1
12	1	1	0	0	1	1	1	0
13	1	1	0	1	1	1	0	0
14	1	1	1	0	1	1	0	1
15	1	1	1	1	1	0	0	1
	1	3	7	15	3	4	4	4

Figure 7.6.6 Four-bit gray codes.

associated with color scanning and printing as well as the inherently long transmission time. One system that is relatively simple in concept transmits two binary colors: black and red. Each color is transmitted independently by using the CCITT Group 3 compression algorithm. The printing process employs thermal paper where the two colors are generated by using two different levels of thermal excitation. Such a system is useful for computer generated graphics with red highlights and for indicating handwritten editorial comments in red.

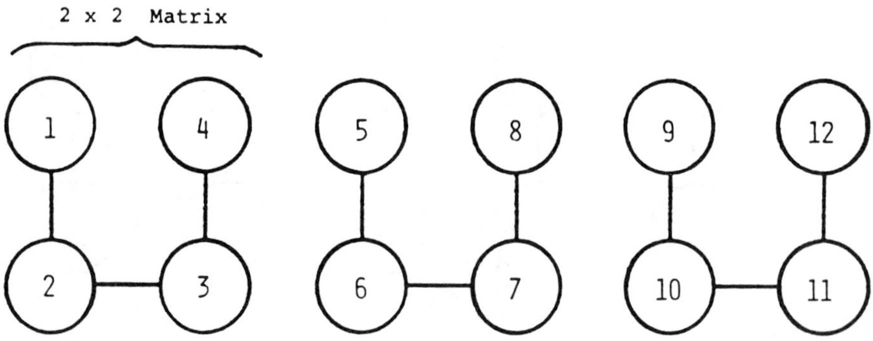

Figure 7.6.7 Pels used for autoresolution.

One color facsimile system scans a Polaroid print on the input and provides another Polaroid print as the output. A drum scanner has three photosensors, one for each primary color. Printing is accomplished by a three-color rotating wheel located in the printing light path. In this case, line sequential color signals are transmitted, and each color is reproduced in a continuous tone rather than a binary image. At the opposite end of the cost scale are color facsimile systems used to transmit newspapers and magazines. *USA Today* is a newspaper which employs color and is transmitted by satellite from a central location to multiple printing stations. In this case, the multiple colors are transmitted and printed in binary form. Compression is employed to minimize the transmission time.

Most of the color facsimile systems developed to date transmit multiple color signals with full resolution. For example, if RGB (red, green, blue) signals were transmitted, the transmission time would be three times that for monochrome-only transmission. Consider the following hypothetical parameters for a continuous-tone color facsimile system:

Resolution	200 pels/inch
Page size	8½ × 11 inches
Transmitted bits/pixel (compressed)	2
Color signals (e.g., RGB)	3
Transmission bit rate	9600 b/s
Transmission time	39 minutes

The long transmission time of 39 minutes points out one of the primary reasons that color facsimile has been introduced slowly into the marketplace. There are several ways that this excessive transmission time can be reduced. The resolution can be reduced with attendant loss of picture quality. A more aggressive compression algorithm could be employed to reduce the transmitted bits/pel from 2 to 1. Transmission time can also be reduced by employing a more expensive data link operating, for example, at 56 kb/s. Finally, to transmit the three color signals with full bandwidth is wasteful because the eye is incapable of perceiving high resolution color. Transmitting a full resolution luminance signal (Y) supplemented by two low resolution chroma signals would be more efficient. This concept is now employed in the two major commercial TV transmission standards. Table 7.7.1 summarizes the relative bandwidths of these standards where B is the luminance bandwidth.

Table 7.7.1. Relative Bandwidths

	NTSC Standard	*PAL Standard (Europe)*
Luminance	(Y) B	B
Chroma	(I) B/2	(U) B/3
Chroma	(Q) B/6	(V) B/3
Total	1.67 B	1.67 B

The reader will note that both TV standards transmit the three signals in the reduced channel capacity of 1.67 B rather than 3 B with a net compression of 1.8. In general, employing the European YUV components for transmission is more convenient than YIQ because the two chroma components have an equal bandwidth.

For the same reason that the *horizontal* chroma resolution can be reduced relative to the luminance without being perceived by the eye, so the *vertical* resolution can, and should, be similarly reduced to effect a bit-rate compression. The simplest way to accomplish this reduction of vertical chroma resolution is to transmit the two chroma signals (U, V) on a line-alternating basis. In this way the total channel capacity can be reduced from 3 B to 1.33 B for a net compression of 2.25. At the receiver, the line chroma signal that was not transmitted is replaced by the signal which was transmitted on the previous line. An additional reason for the lack of development of color facsimile is the absence of standards.

The CCITT is attempting to correct this problem by developing a standard for color facsimile as part of the Group 4 recommendation. It is likely that the first work in this direction will be the standardization of a three-color binary facsimile process, which would permit the printing of full color pages using perhaps the 400-pel/inch resolution. Such a standard would prove advantageous as part of the Office Document Architecture Recommendation, where documents may be computer generated by using electronic publishing technology.

The CCITT is engaged in another standard-setting activity for color imagery which may, or may not, be considered facsimile. The Joint Photographic Experts Group (JPEG) is developing a standard for the compression of continuous-tone color pictures for efficient storage and transmission over ISDN channels. The typical output is anticipated to be a soft copy on a CRT terminal where a usual application would be the rapid retrieval of documents from a remote database. Obviously, the user is assumed able to print out a copy of what he or she is viewing as required. Conventional facsimile systems scan and transmit the page sequentially, pel by pel. The JPEG system employs progressive transmission rather than the sequential process. In progressive transmission, the entire page is first defined with low resolution and reduced picture quality (e.g., 0.1 bits/pel). As time passes the resolution and quality of the page continually improve until the maximum quality is achieved.

This type of progressive transmission and presentation is particularly valuable when the user is browsing page by page through a document. The viewer may rapidly determine, based on the low resolution version of the page, that he or she is not interested in that page, and can quickly request to view the next page. Or, if the viewer is interested, he or she can continue to view the page as the quality improves. The plan is to provide the maximum picture quality in no more than five seconds when transmitted over a 64-kb/s ISDN channel.

The JPEG algorithm will employ the YUV color structure described earlier in this section. The chrominance components are subsampled 2:1 horizontally relative to the luminance component. The compression algorithm will be the discrete cosine transform (DCT) where the image will be divided into blocks of 8×8 pels. The DCT coefficients are transmitted in the conventional zigzag pattern described in Section 7.4. The transform coefficients will be further compressed by using either arithmetic coding or adaptive Huffman coding.

REFERENCES

1. Gersho, A., "On the Structure of Vector Quantizers," *IEEE Transactions on Information Theory,* Vol. IT-28, No. 2, March 1982, pp. 157–166.
2. Linde, Y., Buzo, A., and Gray, R. M., "An Algorithm for Vector Quantizer Design," *IEEE Transactions on Communications,* Vol. COM-28, No. 1, January 1980, pp. 84–95.
3. Gray, R. M., and Linde, Y., "Vector Quantizers and Predictive Quantizers for Gauss-Markov Sources," *IEEE Transactions on Communications,* Vol. COM-30, No. 2, February 1982, pp. 381–389.
4. Huang, H.-M., and Woods, J. W., "Predictive Vector Quantization of Images," *IEEE Transactions on Communications,* Vol. COM-33, No. 11, November 1985, pp. 1208–1219.
5. Contribution of the Japanese Delegation, "Component Vector Quantization," Annex 4 of CCITT Study Group VIII, Geneva, December 1986.

6. Helden, J., and Boekee, D. E., "Vector Quantization Using a Generalized Tree Search Algorithm," *Proceedings Fifth Symposium on Information Theory in the Benelux,* Aalten, May 1984, pp. 21–27.
7. Gersho, A., and Ramamurthi, B., "Image Coding Using Vector Quantization," *Proceedings ICASSP,* Paris, 1982.

Chapter 8
Fax Test Charts

8.1 TEST CHARTS

Test charts made to demonstrate the best recorded copy for sales purposes will not be discussed here. The following test charts are for testing fax equipment performance limitations and transmission channel characteristics.

8.1.1 IEEE Chart 167A

Developing definitions of facsimile terms, test procedures, and test charts has been among the functions of the Institute of Electrical and Electronic Engineers (IEEE) and its predecessor organizations, the Institute of Radio Engineers (IRE) and the American Institute of Electrical Engineers (AIEE), since the 1940s. In the 1960s, the IEEE and the EIA reached an informal agreement regarding facsimile technical activities. IEEE would continue work in areas of definitions of terms, test charts, and performance test procedures, and EIA would pursue equipment and transmission standards for facsimile. The first IRE Facsimile Test Chart was issued in 1955. Decades later, after many updates and reissues, this chart is still widely used for testing performance of fax, electronic imagery devices, office copiers, and other photographic systems. (See Figure 8.1.1.1.)

Chart 167A has a high degree of precision and accuracy of reproduction to allow measurement of equipment performance and transmission channel performance. A full range of continuous tone gray scale is achieved from paper white to a saturated black density of 1.8. The wide variety of test patterns makes it easy to spot transmission problems and loss of resolution. Although some of the patterns may appear to be redundant, each one is better than others for a particular resolution measurement or distortion evaluation. The original chart was assembled from favorite test patterns of each member of the IRE Facsimile Committee. It was expected that after all of the patterns were assembled on one chart, many could be eliminated, but each one was defended and none were eliminated. Patterns 14A and 14B are copies of the National Bureau of Standards (NBS) Microcopy

Figure 8.1.1.1 CCITT facsimile test chart 167A.

Resolution Test Chart without the three lowest resolution patterns. The highest resolution NBS pattern corresponds to 36 lines/mm (914 lines/inch), but the contrast of this pattern is low and the lines are broken. Even the microdefects in some of the patterns are useful for testing. Users are instantly able to spot distortions of the received copy's gray scale by looking at the picture in the test chart.

The Eastman Kodak secretary whose picture appeared on all issues could not have realized that her likeness would be spread worldwide on more than 30,000 facsimile test charts. Each chart as it was used made many more fax copies until her image is universally recognized and associated with facsimile test patterns. Information on the purchase of this chart can be obtained through the IEEE Standards Office in New York.

8.1.2 CCITT Chart No. 1

In 1960 the CCITT issued Recommendation T.20 for Test Chart No. 1. Its size is 4.33 × 9.84 inch (110 × 250 mm). This chart was designed to provide a reliable and rapid means of checking the quality of test transmissions according to uniform principles. The chart allows performance comparisons between different equipment in a precise manner, and quality evaluation of both picture and black-and-white transmissions. A second edition was issued in October 1968. Gray-scale performance is checked by a 15-step density pattern and the photograph. Black-white patterns measure single-line and multiple-line resolution up to 6 lines/mm (152 lines/inch). Other patterns will show analog facsimile distortions caused by telephone lines and equipment design. (See Figure 8.1.2.1.)

This chart was designed for the same purpose as the IEEE Facsimile Test Chart, but has fewer test patterns.

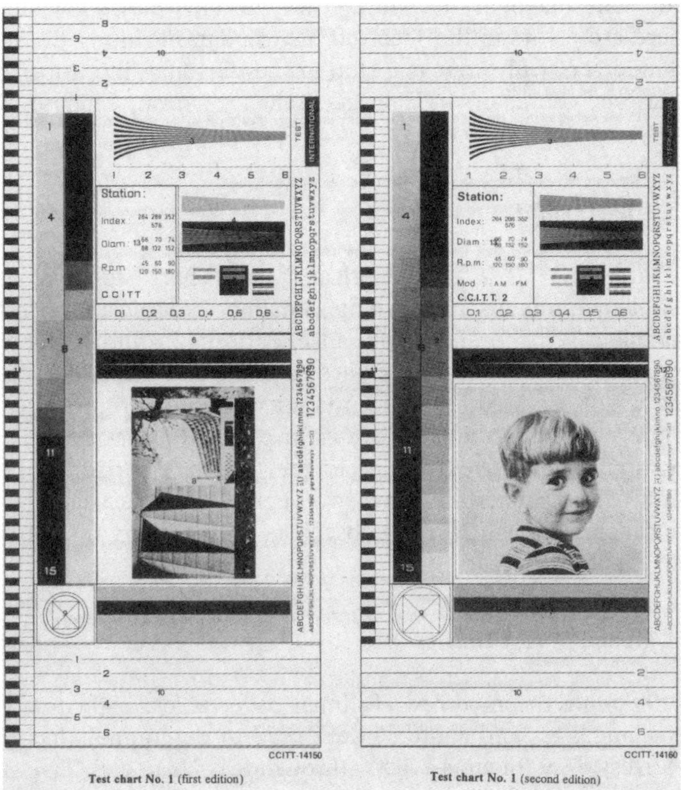

Test chart No. 1 (first edition)   Test chart No. 1 (second edition)

Figure 8.1.2.1 CCITT Test Chart No. 1.

8.1.3 CCITT Chart No. 2

In 1980 the CCITT issued Recommendation T.21 for Test Charts No. 2 and No. 3. Their size is 8.23 × 11.69 inch (210 × 297 mm). These charts for checking the quality of document facsimile transmissions were issued at the same time as the Recommendation T.4 for Group 3 facsimile.

Test Chart No. 2 is a "transmission test chart" for evaluating legibility of text in various languages. It contains test patterns for evaluation of distortion and character groups for evaluation of readability. (See Figure 8.1.3.1.)

The borders are calibrated in millimeters, allowing measurement of the usable area of the received fax copy and errors in centering the page. The test chart size is based on A4 international paper size instead of 8½ × 11 inches as used in North America. The edge-to-edge pattern width is only 8¼ inches, narrower than that provided by some Group 3 fax units. The circle will reproduce as an oblong shape if there is a paper feed rate error. Measurement of the border length will confirm such an error. The diagonal line through the patterns may show steps or white streaks if there are errors in the received picture data. Printing the previous line for MR coding errors will show as a step in the diagonal line. Most of the chart consists of text and characters in English, French, Spanish, Chinese, Arabic, and Russian.

8.1.4 CCITT Chart No. 3

Test Chart No. 3 is a "characterization test chart" for evaluation of the technical quality parameters of fax units, enabling technicians to detect faults, make adjustments, and calibrate performance. Patterns 3.4 and 3.5 show the system performance with thin, isolated, black lines on white and white lines on black. For the same line width, isolated lines (as in type) will usually reproduce better than black-and-white patterns of equal width as in patterns 3.6 through 3.11. Pattern 3.6 has alternating lines corresponding to the Group 3 horizontal resolution of 203/inch (8/mm). These lines are tilted two degrees from vertical, allowing alignment and misalignment with the photosensors as the chart moves through the scanner. (See Figure 8.1.4.1.)

8.1.5 CCITT Group 3 Test Images

When designing the modified Huffman code tables, eight test pages, each with different languages and content, were used to determine which code words should be shortest. (See Figures 8.1.5.1 through 8.1.5.8.)

These images were scanned and a histogram was made to show the number of times that each white run length and each black run length occurred. The

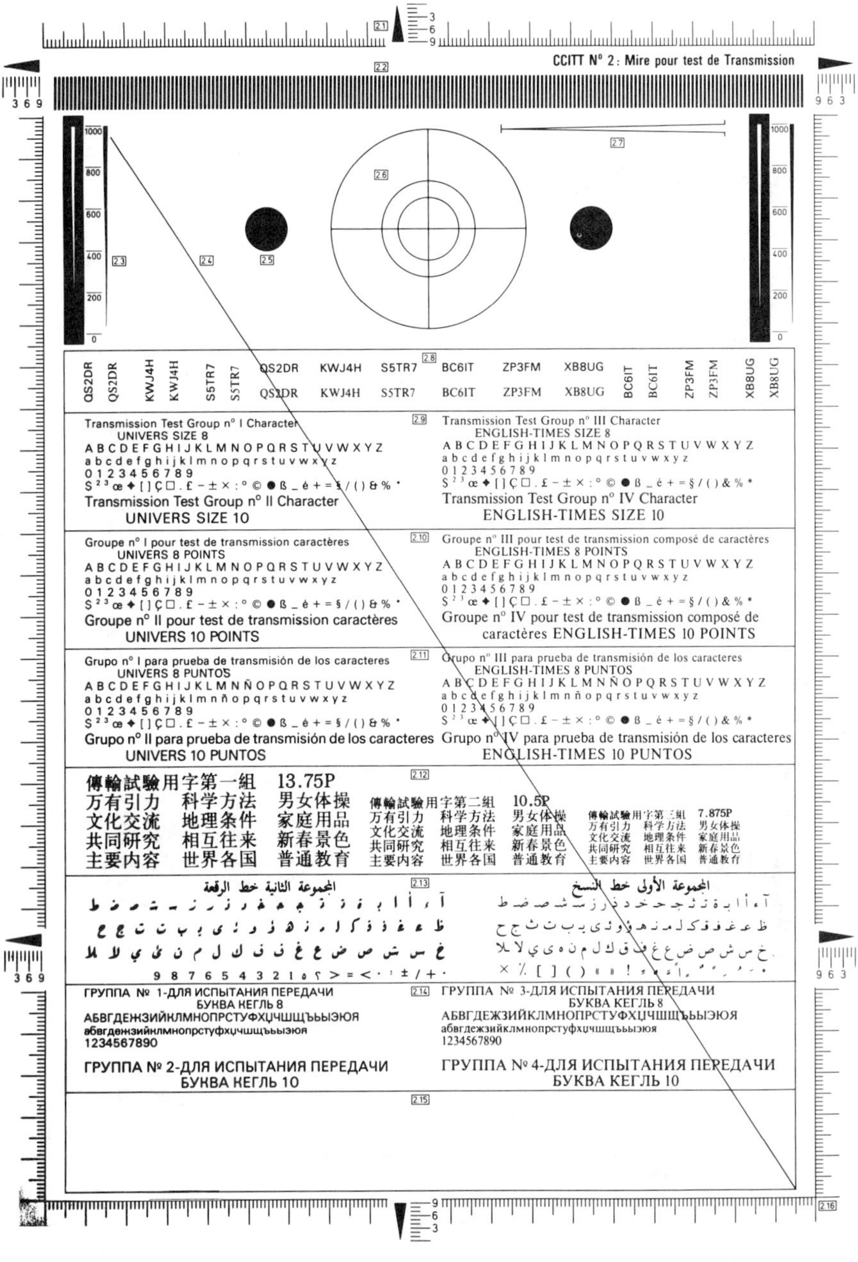

Figure 8.1.3.1 CCITT Test Chart No. 2.

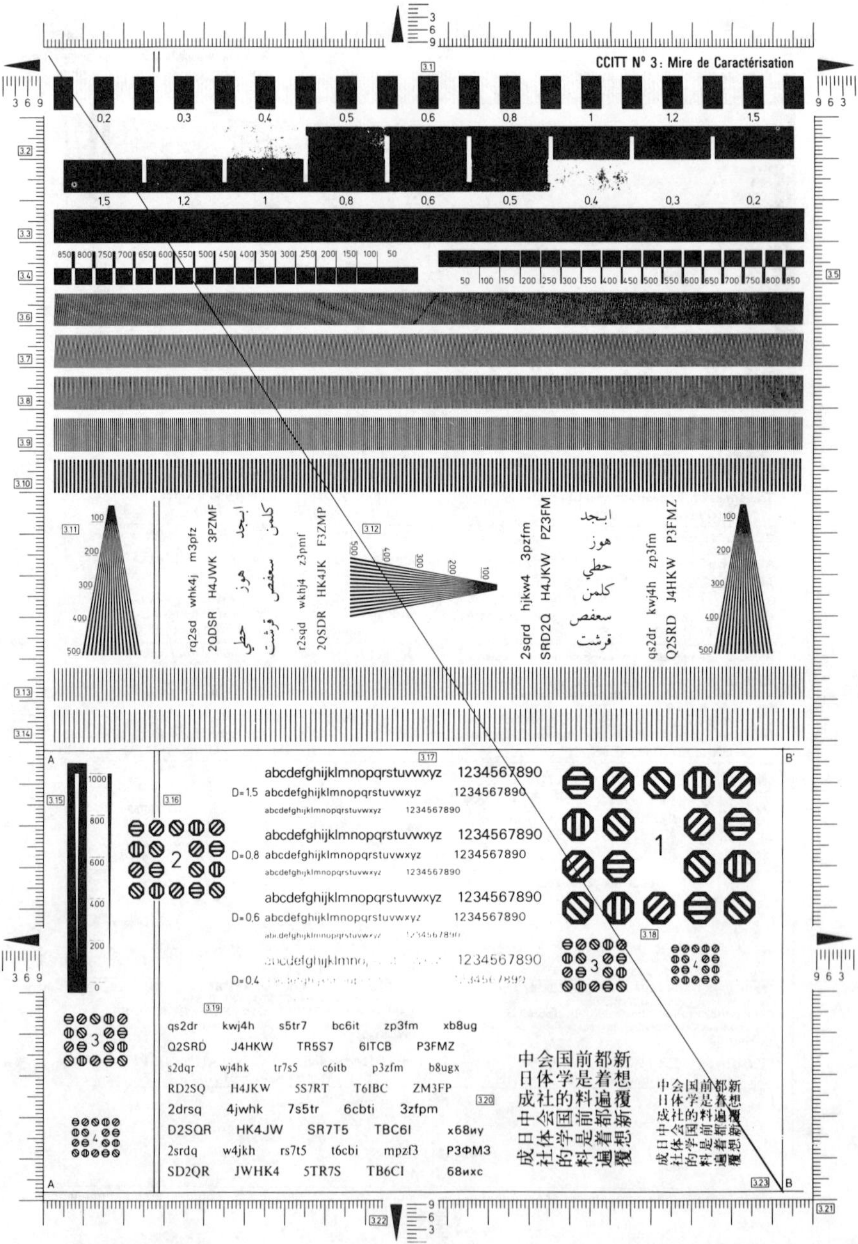

Figure 8.1.4.1 CCITT Test Chart No. 3.

modified Huffman code assigned short code words to those runs that occurred most frequently.

The CCITT does not recognize these test images as official standards and does not have them for sale. They have become *de facto* standards, however. Images Nos. 1, 5, 7, and the Delta Facsimile Test Chart have been scanned at resolutions of 200, 240, 300, 400, and 480 lines per inch and recorded on magnetic tape. One 2400-foot reel of 9-track tape at 1600 dpi in IBM standard EBCDIC format contains these 20 binary images.*

Test Image No. 1, a short business letter, is frequently used by sellers for measuring transmission time for Group 3 fax units. The time listed does not include handshake.

THE SLEREXE COMPANY LIMITED

SAPORS LANE - BOOLE - DORSET - BH 25 8 ER

TELEPHONE BOOLE (945 13) 51617 - TELEX 123456

Our Ref. 350/PJC/EAC 18th January, 1972.

Dr. P.N. Cundall,
Mining Surveys Ltd.,
Holroyd Road,
Reading,
Berks.

Dear Pete,

Permit me to introduce you to the facility of facsimile transmission.

In facsimile a photocell is caused to perform a raster scan over the subject copy. The variations of print density on the document cause the photocell to generate an analogous electrical video signal. This signal is used to modulate a carrier, which is transmitted to a remote destination over a radio or cable communications link.

At the remote terminal, demodulation reconstructs the video signal, which is used to modulate the density of print produced by a printing device. This device is scanning in a raster scan synchronised with that at the transmitting terminal. As a result, a facsimile copy of the subject document is produced.

Probably you have uses for this facility in your organisation.

Yours sincerely,

Phil.

P.J. CROSS
Group Leader - Facsimile Research

Figure 8.1.5.1 CCITT Group 3 test image: (1) Business letter.

* This tape can be obtained from National Communications System, Att: NCS-TS (D. Bodson), Washington, DC 20305.

ÉTABLISSEMENTS ABCDEFG
SOCIÉTÉ ANONYME AU CAPITAL DE 300 000 F
20, RUE DU XVUTRSTSSL F 00000 NTBCLAG
Tél. : (35) 24.46.32 Adr. Tg. : NRVLJROLM
Télex : 31596 F IN : 718490070257
Transporteur (ou Transitaire)
M. M. DUPONT Frères
8 quai des blcdfsh F 0000 NTBCLAG

Mot directeur

CLASSEMENT	FACTURE INVOICE	Exemplaire 15	
CODE CLIENT Z 04599	DATE 7-7-74	NUMÉRO 06	FEUILLET 01

Votre commande du 74-2-2 numéro 438
Notre offre AZ/B7 du 74-1-1 numéro 12

LIVRAISON
5, rue XYZ
99000 VILLE

FACTURATION
12, rue ABCD BP 15
99000 VILLE

DOMICILIATION BANCAIRE DU VENDEUR

CODE BANQUE | CODE GUICHET | COMPTE CLIENT

ORIGINE	TRANSPORTS DESTINATION	MODE
Pays 1	Etat 2	Air

PAYS D'ORIGINE PAYS DE DESTINATION
CONDITIONS DE LIVRAISON DATE 74-03-03
LICENCE D'EXPORTATION NATURE DU CONTRAT (monnaie) FAB
CONDITIONS DE PAIEMENT (échéance, %...)

MARQUES ET NUMÉROS MARKS AND NUMBERS		NOMBRE ET NATURE DES COLIS : DÉNOMINATION DE LA MARCHANDISE NUMBER AND KING OF PACKAGES: DESCRIPTION OF GOODS	NOMEN-CLATURE STATISTICAL No.	MASSE NETTE NET WEIGHT MASSE BRUTE GROSS WEIGHT	VALEUR VALUE DIMENSIONS MEASURE-MENTS
74.21.456.44.2 A		1 Composants	U 123/4	5 kg 8 kg	1400 X 13x10x6
QUANTITÉ COMMANDÉE ET UNITÉ QUANTITY ORDERED AND UNIT	N° ET RÉF. DE L'ARTICLE	DÉSIGNATION	QUANTITÉ LIVRÉE ET UNITÉ QUANTITY DELIVERED AND UNIT	PRIX UNITAIRE UNIT PRICE	MONTANT TOTAL TOTAL AMOUNT
2	AF-809	Circuit intégré	2	104,33 F	208,66 F
10	S8-T4	Connecteur	10	83,10 F	831,00 F
25	ZI07	Composant indéterminé	20	15,00 F	300,00 F

		Inclus	Non inclus
Costs	Débours		
Packing	Emballages		92,14
Freight	Transport		
Insurance	Assurances		
Total invoice amount	Montant total de la facture		1431,80
Installment	Acomptes		
NET TO BE PAID	NET A RÉGLER		1431,80

Figure 8.1.5.2 CCITT Group 3 test image: (2) Invoice.

CCITTの概要

沿革

CCITTは、国際電気通信連合（ITU）の四つの常設機関（事務総局、国際周波数登録委員会、CCIR、CCITT）の一つとして、ITUの中でも、世界の国際通信上の諸問題を真先に取上げ、その解決方法を見出して行く重要な機関である。日本名は、国際電信電話諮問委員会と称する。

CCITTの前身は、CCIF（国際電話諮問委員会）とCCIT（国際電信諮問委員会）である。CCIFは、1924年にヨーロッパに「国際長距離電話通信諮問委員会」が設置され、これが1925年のパリ電信電話会議のとき、正式に、「国際電話諮問委員会」として万国電信連合の公式機関となったものである。CCITは、同じく1925年の会議のとき、CCIFと併立するものとして設置された。

そして、CCIFは、1956年の12月に第18回総会が開催されたのち、CCITは、同年同月に第8回総会が開催されたのち、併合されて現在のCCITTとなった。このCCITTは、CCIFとCCITが解散した直後、第1回総会を開催し、第2回総会は、1960年にニューデリーで、第3回総会は、1964年、ジュネーブで、第4回総会は、1968年、アルゼンチンで開催された。

CCIFとCCITが合併したのは、有線電気通信の分野、とくに伝送路について電信回線と電話回線とを技術的に分ける意味がなくなってきたこと、各国とも大体において、電信部門と電話部門は同一組織内にあること、CCIFの事務局とCCITの事務局の合併による能率増進等がおもな理由であった。

CCITTは、上述のように、ヨーロッパ内の国ぐにによって、ヨーロッパ内の電信・電話の技術・運用・料金の基準を定め、あるいは統一をはかってきたので、現在でも、その影響を受け、会合参加国は、ヨーロッパの国が多く、ヨーロッパで生起する問題の研究が多い。たとえば、1960年のCCITT勧告の中で、技術上配慮する距離は約2,500kmであったが、これはヨーロッパ内領域を想定したものである。

しかしながら、1956年9月に敷設された大西洋横断電話ケーブルは、大陸間電話通信の自動化および半自動化への技術的可能性を与え、CCITTがこの問題を取り上げるに及び、CCITTの性格は漸次、汎世界的色彩を実質的に帯びるに至った。この汎世界的性格は第2次世界大戦後目ざましくなったアジア・アフリカ植民地の独立に伴ってITUの構成員の中にこれらの国が加わり、ITUの中に新しい意見が導入されたことにも起因して、技術面、政治面の双方から導入されてきた。CCITTの汎世界化は、1960年の第2回総会がニューデリーで開催されたことにもあらわれている。この総会までは、CCIT、CCIFのいずれにしろ、アメリカやアジアで総会が開催されたことがなく、CCITT委員長も、ニューデリー総会の準備文書で、この点には注目すべきであるとのべている。

任務

ITUは、全権委員会議、主管庁会議を始めとして、七つの機関をもち、それぞれの機関の権限と任務は国際電気通信条約に明記されている。そこで条約を参照してみるならば、CCITTの任務は、つぎのとおりとなっている。

「国際電信電話諮問委員会(CCITT)は、電信および電話に関する技術、運用および料金の問題について研究し、および意見を表明することを任務とする。」（1965年モントルー条約第187号）

「各国際諮問委員会は、その任務の遂行に当たって、新しい国または発展の途上にある国における地域的および国際的分野にわたる電気通信の創設、発達および改善に直接関連のある問題について研究し、および意見を作成するように妥当な注意を払わなければならない。」（同第188号）

「各国際諮問委員会は、また、関係国の要請に基づき、その国内電気通信の問題について研究し、かつ、勧告を行なうことができる。」（同第189号）

上記第187号と第188号にいわれる「意見」とは、フランス語の Avis から訳したもので、英語では、「勧告(Recommendation)」となっている。CCITTの表明する意見は、国際法的には強制力をもたないものであって、この点が、条約、電信規則、電話規則等各国を拘束する力をもっているものと異なる。もっとも意見とは称しても、技術的分野では、電信規則のごとき、各国政府が承認してその内容を実施する強制規則をもたないので、実際にある機器の仕様を定める場合には、多くの国の意見が統一されたこの「意見」に従わなければ、円滑な国際通信を行なうことができない場合が多い。この意見（または勧告）は、国際通信を行なう場合各国が直面する問題について、具体的意見を表明するもので、たとえば、大陸間ケーブルで大陸間通話を半自動化しようとする場合、その信号方式や取り扱う通話の種類および料金は、どのようにするかを研究して意見を表明する。したがって、CCITTの活動は、つねに時代の最先端を行くもので、CCITTの活動方向は、そのまま世界の国際通信の活動方向であるともいえる。

この意見は、また、電信規則以下のその他の規則のごとく、数年以上の間隔をもって開催される主管庁会議というような大会議の決定をまたなくても表明することができ、また、その改正も容易であるので、現在のように進歩の早い国際通信界では、関係国の意見を統一した国際的見解としては非常に便利である。

Figure 8.1.5.3 CCITT Group 3 test image: (3) Japanese text.

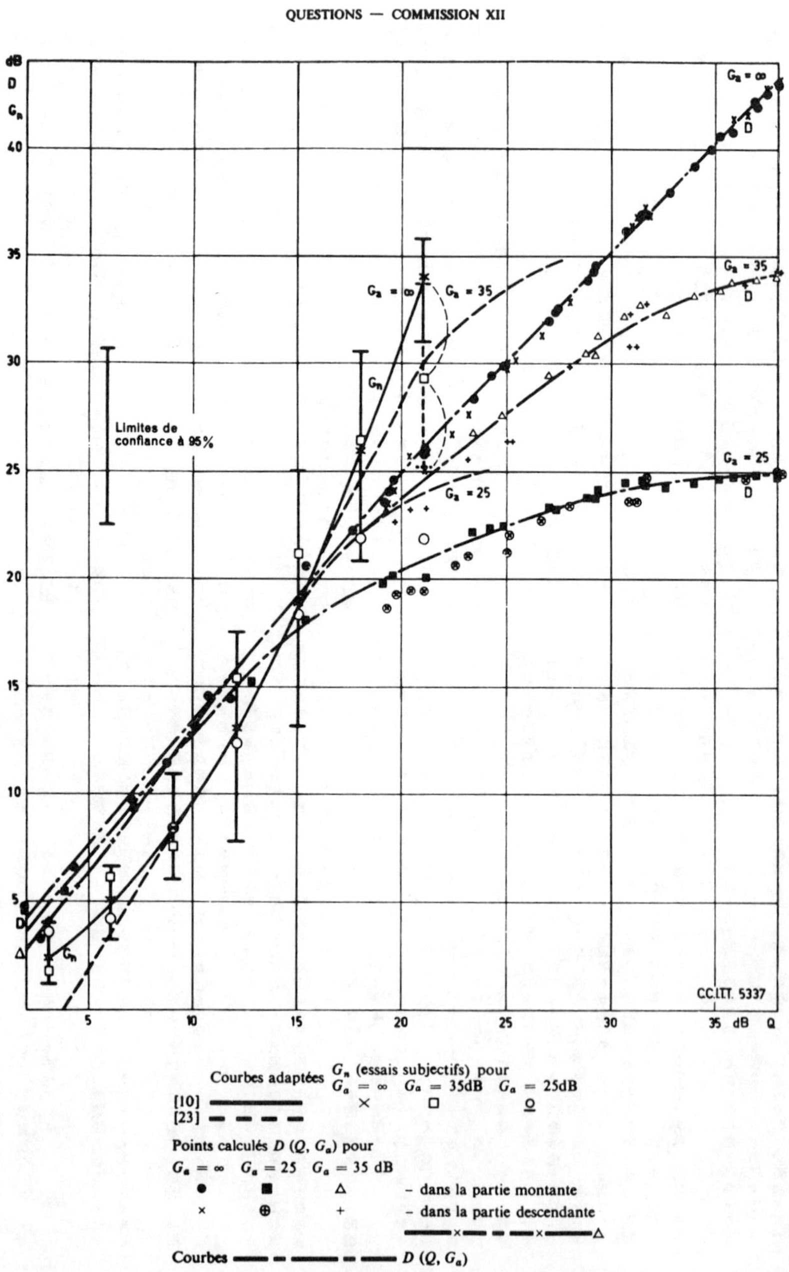

Figure 8.1.5.4 CCITT Group 3 test image: (4) Chart.

L'ordre de lancement et de réalisation des applications fait l'objet de décisions au plus haut niveau de la Direction Générale des Télécommunications. Il n'est certes pas question de construire ce système intégré "en bloc" mais bien au contraire de procéder par étapes, par paliers successifs. Certaines applications, dont la rentabilité ne pourra être assurée, ne seront pas entreprises. Actuellement, sur trente applications qui ont pu être globalement définies, six en sont au stade de l'exploitation, six autres se sont vu donner la priorité pour leur réalisation.
Chaque application est confiée à un "chef de projet", responsable successivement de sa conception, de son analyse-programmation et de sa mise en oeuvre dans une région-pilote. La généralisation ultérieure de l'application réalisée dans cette région-pilote dépend des résultats obtenus et fait l'objet d'une décision de la Direction Générale. Néanmoins, le chef de projet doit dès le départ considérer que son activité a une vocation nationale donc refuser tout particularisme régional. Il est aidé d'une équipe d'analystes-programmeurs et entouré d'un "groupe de conception" chargé de rédiger le document de "définition des objectifs globaux" puis le "cahier des charges" de l'application, qui sont adressés pour avis à tous les services utilisateurs potentiels et aux chefs de projet des autres applications. Le groupe de conception comprend 6 à 10 personnes représentant les services les plus divers concernés par le projet,et comporte obligatoirement un bon analyste attaché à l'application.

II - L'IMPLANTATION GEOGRAPHIQUE D'UN RESEAU INFORMATIQUE PERFORMANT

L'organisation de l'entreprise française des télécommunications repose sur l'existence de 20 régions. Des calculateurs ont été implantés dans le passé au moins dans toutes les plus importantes. On trouve ainsi des machines Bull Gamma 30 à Lyon et Marseille, des GE 425 à Lille, Bordeaux, Toulouse et Montpellier, un GE 437 à Massy, enfin quelques machines Bull 300 TI à programmes câblés étaient récemment ou sont encore en service dans les régions de Nancy, Nantes, Limoges, Poitiers et Rouen ; ce parc est essentiellement utilisé pour la comptabilité téléphonique.
A l'avenir, si la plupart des fichiers nécessaires aux applications décrites plus haut peuvent être gérés en temps différé, un certain nombre d'entre eux devront nécessairement être accessibles, voire mis à jour en temps réel : parmi ces derniers le fichier commercial des abonnés, le fichier des renseignements, le fichier des circuits, le fichier technique des abonnés contiendront des quantités considérables d'informations.
Le volume total de caractères à gérer en phase finale sur un ordinateur ayant en charge quelques 500 000 abonnés a été estimé à un milliard de caractères au moins. Au moins le tiers des données seront concernées par des traitements en temps réel.
Aucun des calculateurs énumérés plus haut ne permettait d'envisager de tels traitements.
L'intégration progressive de toutes les applications suppose la création d'un support commun pour toutes les informations, une véritable "Banque de données", répartie sur des moyens de traitement nationaux et régionaux, et qui devra rester alimentée, mise à jour en permanence, à partir de la base de l'entreprise, c'est-à-dire les chantiers, les magasins, les guichets des services d'abonnement, les services de personnel etc.
L'étude des différents fichiers à constituer a donc permis de définir les principales caractéristiques du réseau d'ordinateurs nouveaux à mettre en place pour aborder la réalisation du système informatif. L'obligation de faire appel à des ordinateurs de troisième génération, très puissants et dotés de volumineuses mémoires de masse, a conduit à en réduire substantiellement le nombre.
L'implantation de sept centres de calcul interrégionaux constituera un compromis entre : d'une part le désir de réduire le coût économique de l'ensemble, de faciliter la coordination des équipes d'informaticiens; et d'autre part le refus de créer des centres trop importants difficiles à gérer et à diriger,et posant des problèmes délicats de sécurité. Le regroupement des traitements relatifs à plusieurs régions sur chacun de ces sept centres permettra de leur donner une taille relativement homogène. Chaque centre "gérera" environ un million d'abonnés à la fin du VIème Plan.
La mise en place de ces centres a débuté au début de l'année 1971 : un ordinateur IRIS 50 de la Compagnie Internationale pour l'Informatique a été installé à Toulouse en février ; la même machine vient d'être mise en service au centre de calcul interrégional de Bordeaux.

Figure 8.1.5.5 CCITT Group 3 test image: (5) Very dense document.

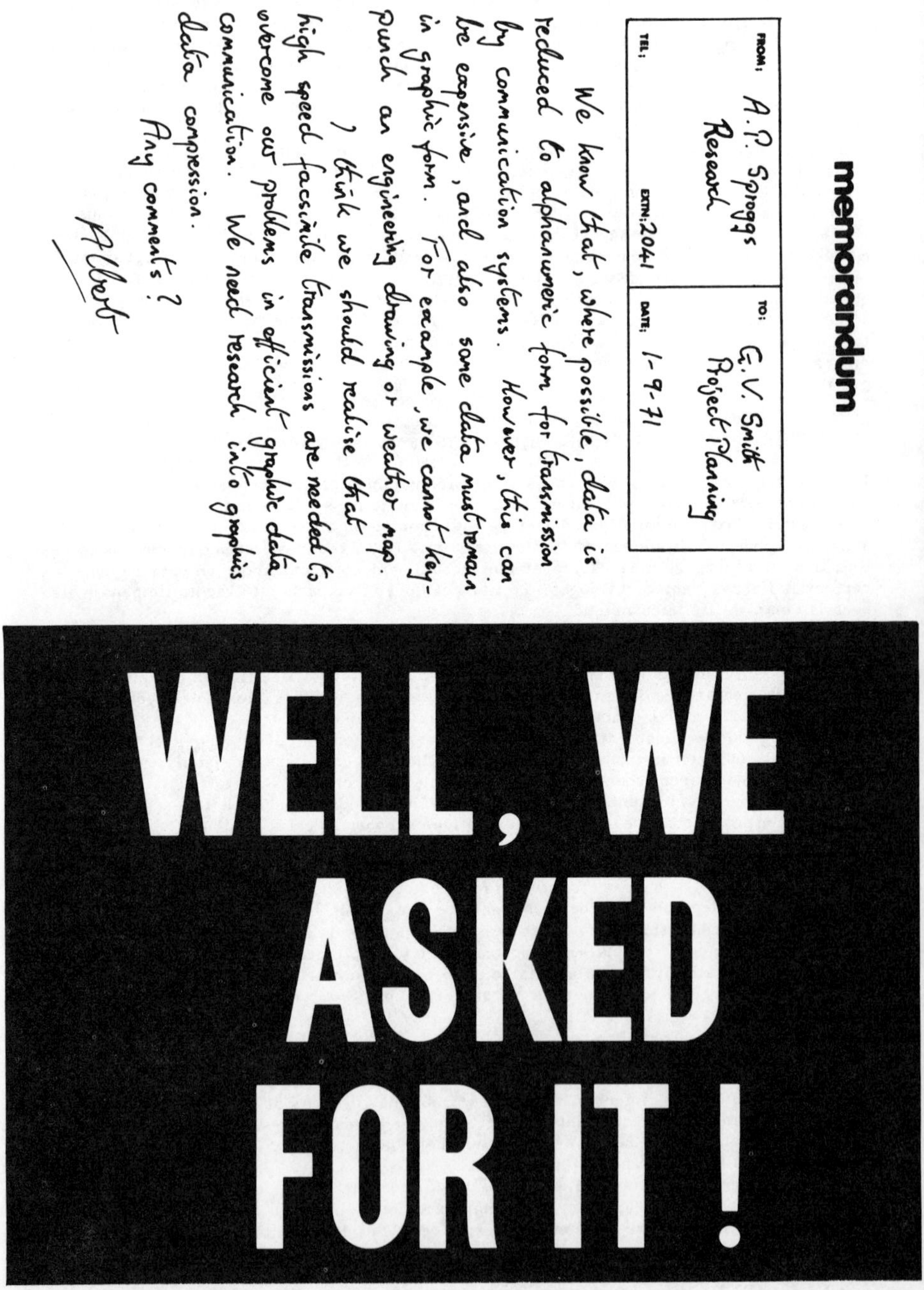

Figure 8.1.5.6 CCITT Group 3 test image: (6) Handwritten note on printed memorandum sheet.

Cela est d'autant plus valable que $T\Delta f$ est plus grand. A cet égard la figure 2 représente la vraie courbe donnant $|\phi(f)|$ en fonction de f pour les valeurs numériques indiquées page précédente.

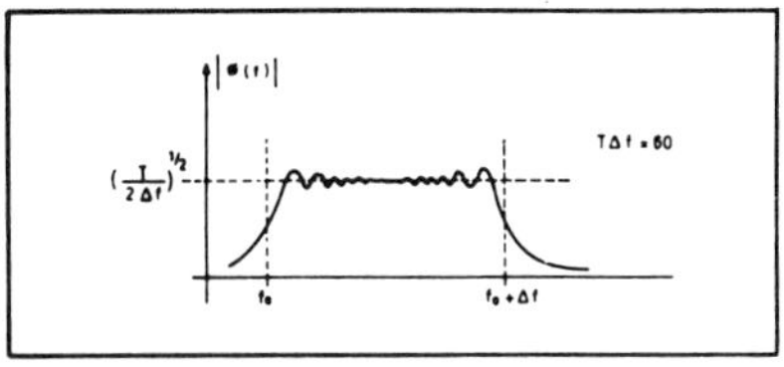

FIG. 2

Dans ce cas, le filtre adapté pourra être constitué, conformément à la figure 3, par la cascade :

— d'un filtre passe-bande de transfert unité pour $f_0 \leqslant f \leqslant f_0 + \Delta f$ et de transfert quasi nul pour $f < f_0$ et $f > f_0 + \Delta f$, filtre ne modifiant pas la phase des composants le traversant ;

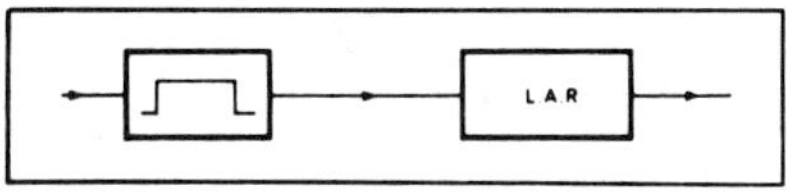

FIG. 3

— filtre suivi d'une ligne à retard (LAR) dispersive ayant un temps de propagation de groupe T_R décroissant linéairement avec la fréquence f suivant l'expression :

$$T_R = T_0 + (f_0 - f)\frac{T}{\Delta f} \quad (\text{avec } T_0 > T)$$

(voir fig. 4).

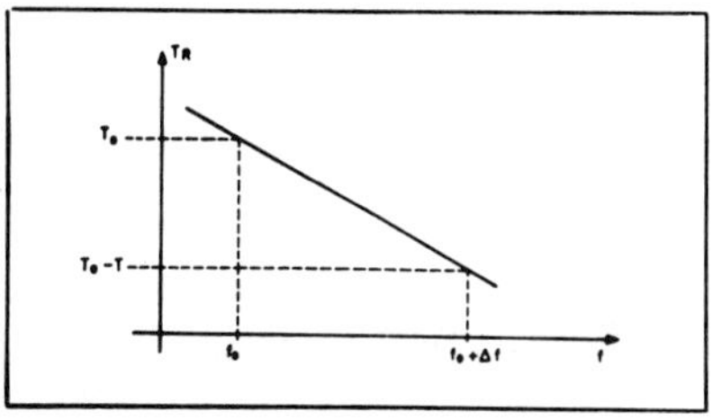

FIG. 4

telle ligne à retard est donnée par :

$$\varphi = -2\pi \int_0^f T_R \, \mathrm{d}f$$

$$\varphi = -2\pi \left[T_0 + \frac{f_0 T}{\Delta f} \right] f + \pi \frac{T}{\Delta f} f^2$$

Et cette phase est bien l'opposé de $\underline{/\phi(f)}$,

à un déphasage constant près (sans importance) et à un retard T_0 près (inévitable).

Un signal utile $S(t)$ traversant un tel filtre adapté donne à la sortie (à un retard T_0 près et à un déphasage près de la porteuse) un signal dont la transformée de Fourier est réelle, constante entre f_0 et $f_0 + \Delta f$, et nulle de part et d'autre de f_0 et de $f_0 + \Delta f$, c'est-à-dire un signal de fréquence porteuse $f_0 + \Delta f/2$ et dont l'enveloppe a la forme indiquée à la figure 5, où l'on a représenté simultanément le signal $S(t)$ et le signal $S_1(t)$ correspondant obtenu à la sortie du filtre adapté. On comprend le nom de récepteur à compression d'impulsion donné à ce genre de filtre adapté : la « largeur » (à 3 dB) du signal comprimé étant égale à $1/\Delta f$, le rapport de compression est de $\dfrac{T}{1/\Delta f} = T\Delta f$

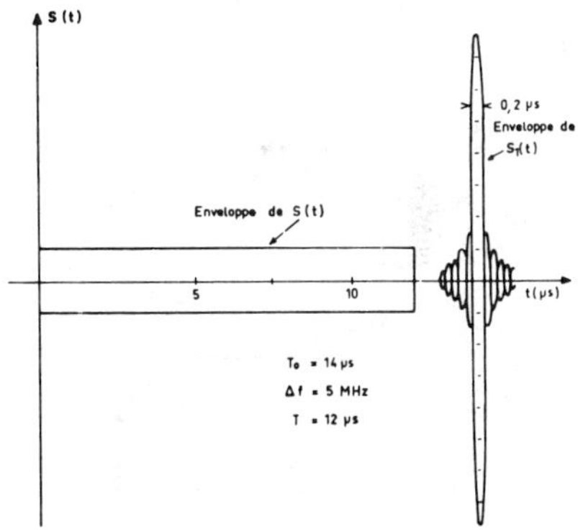

FIG. 5

On saisit physiquement le phénomène de compression en réalisant que lorsque le signal $S(t)$ entre dans la ligne à retard (LAR) la fréquence qui entre la première à l'instant 0 est la fréquence basse f_0, qui met un temps T_0 pour traverser. La fréquence f entre à l'instant $t = (f - f_0)\dfrac{T}{\Delta f}$ et elle met un temps $T_0 - (f - f_0)\dfrac{T}{\Delta f}$ pour traverser, ce qui la fait ressortir à l'instant T_0 également. Ainsi donc, le signal $S(t)$

Figure 8.1.5.7 CCITT Group 3 test image: (7) Printed technical drawings and mathematical equations and notation.

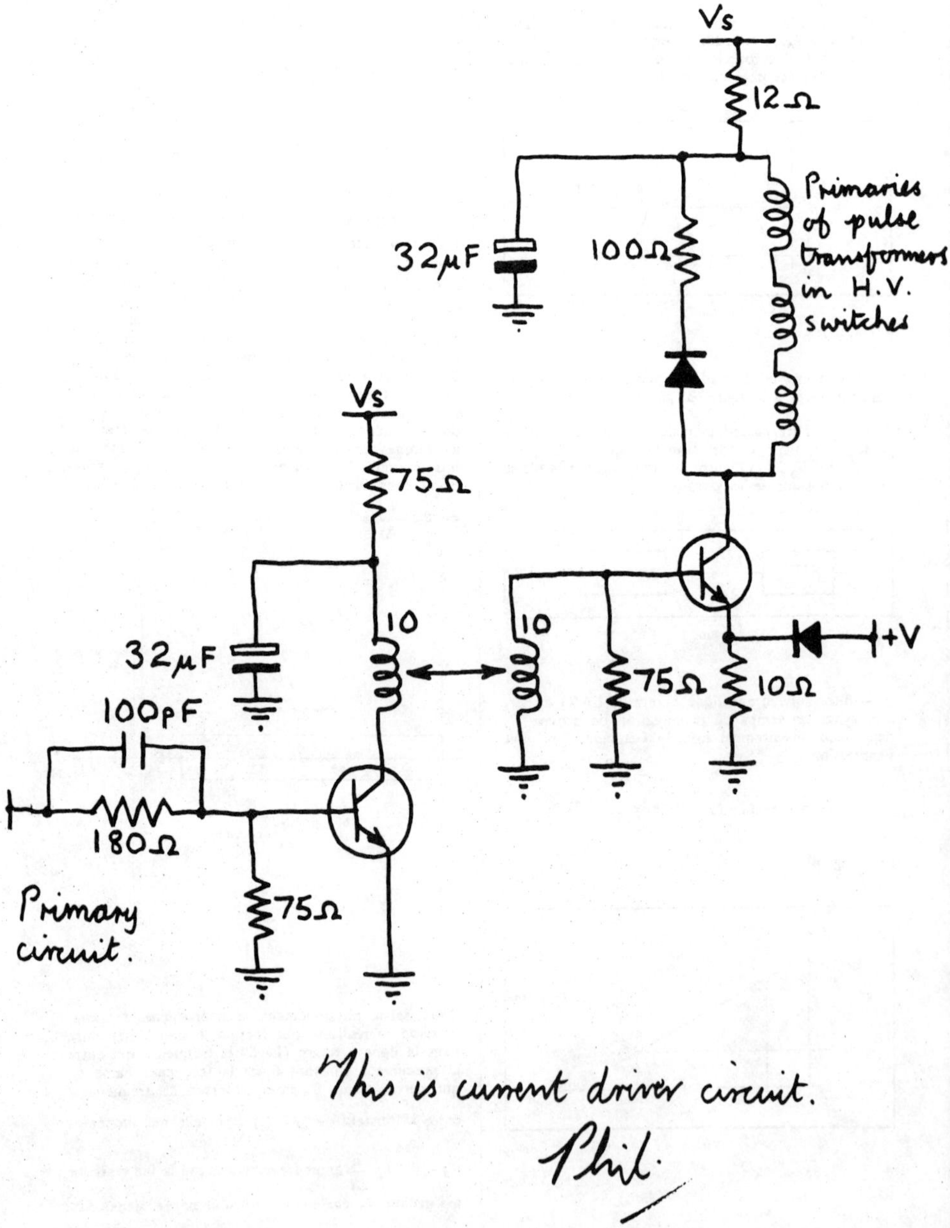

Figure 8.1.5.8 CCITT Group 3 test image: (8) Pencil sketch and handwritten comment.

Chapter 9
The Marketplace

9.1 THE JAPANESE FAX MONOPOLY

Japanese dominate the facsimile market, supplying about 90% of the world's total production. Exports from Japan doubled in 1988 as demand for fax units soared in Europe and the United States. Production costs have risen in Japan so that the market is opening up to other countries with low labor rates, and production has also started in the countries that are large fax users. Direct exports from Japan are not expected to grow as fast in the future as production shifts to other countries. Most of this overseas production is controlled by Japanese companies which have the technical and manufacturing expertise. There are more than twenty Japanese facsimile manufacturers, but five of them account for most of the production. (See Table 9.1.1.)

Table 9.1.1. Japanese Fax Production
(1987 = 364 billion yen)

Company	*Percent*
Ricoh	21.0
Matsushita	20.5
Canon	13.0
NEC	13.0
Toshiba	13.0
Others	19.5

All of the leading Japanese fax producers are already manufacturers of other electronic equipment in countries where they have large markets. The five biggest manufacturers have started or will soon start fax production overseas. Canon produces fax in France and plans to start production in Virginia soon. Matsushita is manufacturing in Singapore for export to the United States. Nippon Electric Company (NEC) makes fax units in Oregon and plans to make fax units in the United Kingdom. Ricoh manufactures fax units in the United Kingdom and the United States, with France to be added soon. Toshiba also plans to manufacture fax units in France.

As an indicator of the newly found importance of fax, compare the numbers of Japanese fax units produced to the numbers of Japanese office copiers produced for the last few years. (See Table 9.1.2.)

Table 9.1.2. Fax *versus* Copier Production in Japan

	—Produced—		*—Exported—*	
Year	*Fax Units*	*Copiers*	*Fax Units*	*Copiers*
1981	127,000	1,404,000	38,000	
1982	229,000	1,420,000	66,000	
1983	304,000	1,924,000	88,000	1,442,000
1984	511,000	2,340,000	157,000	1,793,000
1985	866,000	2,701,000	338,000	2,152,000
1986	1,234,000	2,580,000	558,000	1,989,000
1987	2,405,000	2,210,000	1,495,000	1,695,000
1988 *	4,200,000	1,940,000	3,600,000	1,600,000

* Estimated.

Notice that twice as many office copiers as fax units were produced in 1986. In 1987, the number of office copiers dropped about 10% as the number of fax units doubled so that 200,000 more were produced than office copiers! The 1988 estimates show manufacture of twice as many fax units as office copiers! The copier market is falling off in Japan and the fax market is zooming. As a result, many Japanese companies have recently started manufacturing fax units.

9.2 THE DEVELOPMENT OF FAX MARKETS

Office facsimile has been around for a long time, but it took major advances in technology to achieve office acceptability. Critical factors were high reliability, speed, good copy quality, and low pricing. Fax only recently has grown to the point where businesses are expected to have one.

For the first 100 years, there was no real fax market. There were more people working on fax development than the number of machines used for more than demonstrations. Table 9.2.1. shows significant items in the development of fax-markets.

Table 9.2.1. Development of Fax Markets

First 100 Years	No Market
1941	Newsphotos — Manufacturers Sell to Press Associations
1943	Photos-Messages Sell to US Government
1965	Magnavox Telecopier Xerox Marketing 25,000 Ordered
1975	100,000 Fax in USA
1980	Group 3 Standard Company Salesmen Later — Retail Sales Channels
1987	Critical Mass Reached Fax Sales Explode!!! Seven Million Group 3 Fax Now

In 1941, the small number of fax units made were for news photo use. Sales were made by press associations like Associated Press or United Press, and newspapers or chains themselves.

In 1943, Order No. 1041-MPD-43 for 300 fax units was placed with Times Facsimile Corporation by the US Army Signal Corps for Facsimile Transceiver FX-1B, part of Facsimile Equipment RC-120. For receiving photos on film or photographic paper, the FX-1B could be operated in a portable "dark tent" with light-tight arm holes. The film was loaded on the drum, and the controls were operated without seeing what you were doing. When a second source did not meet performance standards, follow-on orders were obtained with little marketing effort.

In 1948, Western Union had 40,000 to 50,000 Deskfax™ units in service. The telegram-size page was wrapped around a drum and held there by rolling metal "garter springs" over it. Deskfax™ was designed and manufactured by Western Union and furnished to their telegram service customers.

In 1965, Xerox Corporation placed an order with Magnavox for 25,000 telecopier units with an option for 25,000 more. This started a substantial marketing effort for sales of fax to business customers. Loading a page for sending or receiving on the telecopier was as easy as opening and closing a door. The telephone handset was placed on a small box to match its microphone and speaker connected to the fax unit. The fax unit now talked and listened to fax tones on the telephone line (called *acoustic coupling*). This made the regular telephone line usable for office fax units. A wired-in connection by the Bell System (Data Communication Terminal Equipment 602) was costly, impractical, and little used.

In 1975, after many other US manufacturers increased their sales efforts, 100,000 fax units had been sold. Marketing and sales were usually done directly by the manufacturer. Some joint-venture companies were set up and later bought by the Japanese, who then switched to their own product names for selling.

In 1980, manufacturing was shifting rapidly to Japan, and selling was still mainly through company sales organizations. In Europe and North America, the marketing of Japanese-produced facsimile units was often through original equipment manufacturing (OEM) channels using companies that had earlier designed and manufactured their own fax units.

In 1987, sales started to escalate. Selling channels expanded, adding distributors, dealers, department stores, discount houses, and catalog merchandisers. More than half of the US Group 3 fax units were sold by the National Office Machine Dealers Association (NOMDA) members. Direct sales accounted for one-third, and the rest were sold through distributors and wholesalers. In Western European countries, fax equipment was sold mainly by the PTTs, the government agencies responsible for the post office, telegraph, and telephone services. Selling became easier as businessmen realized that they needed fax units. Japanese companies now sell fax through their own European and American companies.

Major US OEM companies selling fax units are AT&T, Harris-3M, Pitney Bowes, Telautograph, and Xerox. OEM sellers in the United States often compete for sales directly with the Japanese supplier.

The January 1989 issue of *IEEE Spectrum* is called "Technology '89." Gary Stix, Associate Editor writes:

> While high-performance communication technology has only gradually taken hold, it was a long-familiar but freshly packaged technology that hit the big time, catching the fancy of the general public as nothing else since the IBM PC's introduction. Facsimile machines have achieved a new popularity, having in little more than a year doubled in number to 1.2 million in the United States. [See Table 9.2.2.]

In 1988 the number of Group 3 units in service increased by 81% in only one year for North America and increased by 156% for Western Europe. The worldwide growth of Group 3 fax units was 75%. With growth rates like this many reports on fax units in service are very low because the data may be six months to one year old. The Group 4 fax units shown for North America are mostly the Group 4-like fax units bought by Federal Express for their now defunct Zap Mail electronic courier service.

The penetration of fax units into the business environment is indicated by the percentage of business lines that have a fax unit connected. In Japan, almost 20% of business lines are for fax, in the United States, 4%, and in Europe, 2%. If we compare the penetration percentage between the two countries, the United States should be able to sell five times as many fax units as Japan in the near future. This neglects the differences in language, however. Fax in Japan is even more of a business necessity than in the United States. The 4000-plus Japanese ideograms are not so easy to send in ASCII code as is the 26-letter Roman alphabet used in the United States.

The latest models are desk top units, some taking little more space than a modern office telephone. User-friendly automatic operation has been responsible for their huge success in record communication. The lowest priced Group 3 units in 1988 cost about 5% of those introduced at the end of 1980!

Sending documents by fax is surely the fastest way and is usually the cheapest as well. Because information is transferred faster and unnecessary talking eliminated, fax is more economical than a telephone call. Before a courier service could come to pick up a document, fax delivery has been completed.

Delivery costs for fax start at 10 cents or less for local calls. Table 9.2.3 compares Group 3 daytime document delivery costs with courier services and telex.

In the United States, coast to coast fax costs about 60 cents for two pages during business hours, or less than a dollar for five pages. Compare this cost with courier ($14) or postal express service ($8.75), and these services both require the user to drop-off the document for sending. Night fax, at rates less than half daytime,

Table 9.2.2. World Fax Units Installed

Facsimile Equipment Population [a]

(As of December 31, 1987)

	United States and Canada	*Western Europe*	*South America, Africa, and the Middle East*	*Far East*	*Eastern Europe and the USSR*	*Totals*
Group 1	15,000	10,000	9,000	6,000	12,000	52,000
Group 2	100,000	26,000	9,500	65,000	19,000	219,500
Group 3	942,000	660,000	54,000	2,600,000	51,000	4,307,000
Group 4	10,000	250	30	9,000	—	19,280
Totals	1,067,000	696,250	72,530	2,680,000	82,000	4,597,780

Facsimile Equipment Population [a]

(Estimate for December 31, 1988)

	United States and Canada	*Western Europe*	*South America, Africa, and the Middle East*	*Far East*	*Eastern Europe and the USSR*	*Totals*
Group 1	4,000	2,000	7,000	2,000	11,000	26,000
Group 2	53,000	13,000	7,000	20,000	17,000	110,000
Group 3	1,707,000	1,690,000	184,000	3,790,000	151,000	7,522,000
Group 4	15,000	2,000	100	48,000	—	65,100
Totals	1,779,000	1,707,000	198,100	3,860,000	179,000	7,723,100

[a] [9/88] Courtesy of George M. Stamps.

still delivers before couriers. The prime advantage of fax, however, is "immediate delivery."

Although the costs of sending fax compared with mail or courier are most advantageous for a few pages, sending even a 25-page document may be cheaper by fax. Courier service between Paris and London is FFr 250, or $40. Fax transmission costs only FFr 4 ($0.64) for one page, FFr 18 ($2.87) for five pages, or

Table 9.2.3. Document Delivery Costs
(Five Pages from Boston or New York at Daytime Rates)

Service	*Local (Dollars)*	*Chicago (Dollars)*	*West Coast (Dollars)*
Federal Express	14.00	14.00	14.00
Express Mail	8.75	8.75	8.75
United Parcel Service	8.50	8.50	8.50
Telex	17.70	17.70	17.70
Group 3 Fax [a]	0.10	0.66	0.72

[a] *Note:* Fax costs vary with telephone service and document content.

FFr 95 ($15) for 25 pages. Between Nice, France, and Manchester, England, courier service costs FFr 350 ($56). Fax costs are the same as between Paris and London.

9.3 GROUP 3 EQUIPMENT IN THE MARKETPLACE

9.3.1 Hints to Fax Buyers

Features — If you are buying a fax unit, knowing the useful features can save you money. When dealing with sales personnel, it will be to your advantage to know which features will be most helpful for your particular needs. The fax market has been flooded with new models. Some of the low-cost models may not have features needed to satisfy you. Sales personnel may promote many features you do not, and will probably never, need. With higher priced models, you will also be paying for features you will never use. The trick is to find which features you really want and to ignore the others.

Reliability — Reliability is generally very good, with most fax units not requiring service for one to two years or more. Plain paper fax units have more mechanical parts which may malfunction, reducing reliability somewhat.

Warranty — Warranty varies from ninety days to two years.

Service Contract — Before signing a service contract, establish the guaranteed maximum response-time offered.

Installation and Training — Installation and training might be handled by someone in your organization if you are trying to save money. Some dealers include these items in the purchase price for the fax unit. If possible, plan on using a separate phone line for your fax unit.

Prices — Discounts are available on most fax units. The lowest prices often are from sellers who are not authorized dealers, offer no help in how to set up the equipment, and sell gray market goods with no factory warranty.

Size of Page — Most users need capability for handling full size 8½ × 11-inch or A4-size pages. The lowest price may be for a small page size Group 3 fax unit. Make sure you do not buy an A5, ½-page size, or A6, ¼-page size, unless that is all you want. Copies of documents sent from the standard size A4 units and recorded on A5 units are reduced in size, but usually can still be read if the original print is no smaller than 12 point or 10 pitch (10 characters per inch typewriter size). The Group 3 fax units with A6 half-width recording paper are of very limited use when operating with standard page size fax units. An A6 fax unit is most useful when operating with another A6 unit in a special application where small page size is acceptable. (See Section 2.8.6 for more information on A5 and A6 fax units.)

Paper Roll Size — Most low-end units have a 60 to 98 foot long roll of recording paper. If this length is too short for the number of copies that you expect to receive between convenient changing times, look for a model with 148 to 164 foot intermediate length or a standard featured fax unit with a full 328 foot roll length. Paper curl near the end of short rolls may be annoying.

Paper Cutter — Do not expect the low-end fax unit to have an automatic paper cutter. When a number of pages are received there will be a banner of paper several feet long waiting to be cut by hand. You will be trading low price for the convenience of having the recorded pages cut to length.

Resolution — Fine resolution of 203 lines/inch × 198 lines/inch is usually available on even the lowest priced fax units. It provides better clarity of fax copies for pages with fine print and it is worth the effort to find.

Seconds per Page — Most suppliers quote the time taken for CCITT Test Image No. 1 (Figure 8.1.2.1) sent in standard resolution. Pages with more lines of printing or more detail take more time to send. Modem speed, minimum scan line time, MH or MR coding method, and other factors affect the sending time. The time quoted will exclude the automatic handshake time of about 15 seconds before the first page is sent. Some fax units have a higher speed proprietary mode of operation under the NSF option that reduces the time per page and may reduce the handshake time. This speed-up works only when sending between similar units from the same manufacturer, but will reduce back to standard mode when working with another

Group 3 fax. It may take more time to receive a page than to send one. Time per page listed varies from 9 to 45 seconds for Test Image No. 1. Low-end fax units usually take longer than a standard fax unit to send or receive a page.

Automatic Document Feeder — A few Group 3 fax units selling for less than $1000 have an automatic document feeder for five pages, but most do not. A document feeder can be a time saver if you send more than one page at a time.

RS-232 Digital Connection — This is a feature that very few users need. Users who call other fax units on the regular telephone network (PSTN) have no need for it. Unless you are sure you need it, do not pay extra for it.

9.3.2 Hints to Fax Users

Phone Line for Fax — A separate phone line terminated in an RJ-11 jack and with its own number is usually best. This line should not be a switchboard extension. The operator who completes a switchboard call will not be around after working hours. One of the fax advantages is 24-hour-a-day availability for sending documents to the fax unit. The separate line also cuts down on fax and phone competing with each other. Although they both use the telephone network, fax and phone are separate services. When sending a document of many pages the user should not have to wait before making or receiving important phone calls. If you cannot afford two phone lines, consider buying one of the integrated fax-phone units described in Section 9.3.5 or a fax-phone switch box. The switch box plugs into your telephone jack. Both the telephone and the fax unit plug into jacks on the switch box. When a call is received, the switch box connects the line to either the fax or the phone. (See Section 2.8.9.)

Power — A separate power circuit is recommended for the fax unit. A separate circuit prevents another power user from tripping the circuit breaker and putting the fax out of operation. Leave the power turned on for the fax unit 24 hours a day, 7 days a week, so that it is always ready to receive. This allows use across time zones or when the other party is working late. Standby power is usually about 15 W. A plug-in power line filter, like those sold for computers, is a good investment. You might consider turning the power off during severe thunderstorms.

Location of Fax — Keep the fax unit in your own office or your secretary's to give you direct fax communication without mail-room delays. This also prevents your received documents from being on view to everyone who walks by. Having your own fax saves the time of finding a unit which is not in use or learning why you still have not received documents which were sent to you. Having your fax unit in a separate room or in a message center wastes time. Consider upgrading to a fax model that has a built-in memory mailbox with a security number if confidential documents are to be received.

Overseas Calls — Take advantage of the time zone difference. A fax sent at the end of your workday can generate a reply which will be waiting for you the next morning.

Fax Number Listing — Use your fax number on your business card and in your advertising material to generate additional business. These listings will be seen by your existing and potential customers. A listing in a fax directory can bring additional desirable business contacts, but it may also bring junk fax mail. (See Section 9.3.11.)

Polling — When leaving documents for sending to someone else while you are not there, use a polling code. Otherwise, they could be sent to the wrong party.

Paper Supply — Check when leaving at night to see that there is enough recording paper for overnight fax reception.

Buying Thermal Recording Paper — WARNING: Use of the wrong thermal recording paper can gum up the thermal head so it will not work. This could even cause permanent damage. Do not use the cheapest paper unless the fax manufacturer approves its use. There are low-temperature papers made for low power, small recorders. Used in a standard fax unit, they surely would cause problems.

Cleaning the Fax Unit — Kits are available for the occasional cleaning your fax unit may need. If the recording has streaks, the print head can be gently cleaned with alcohol. If the automatic document feeder sometimes fails to pull in the page being sent, the scanner section paper rollers probably need cleaning. The kit has a simple method of doing this without opening the fax unit.

Fine Mode — Use this higher resolution mode when sending small type, the received fax copy will be used in a report, or you want to make a good impression.

Automatic Document Feeder — Use this feature to send more than one page when you do not want to waste time waiting for each page to be sent. If the pages you want to send are on onion skin paper or dog-eared, make an office copy to send, thus avoiding the possibility of jamming your original in the fax unit. Some fax units have a five-page document feeder. To avoid a paper jam, do not put in more sheets or thick sheets.

Automatic Dialer — Enter the numbers you fax for individuals and groups to whom you will send the same message. This is a time saver with which you do not get wrong numbers.

Originals — If your originals do not feed through the fax unit properly, try to avoid sending pages right after they are made on an office copier. This problem usually happens with a high-usage copier. A volatile coating on the paper used in the copier may be the culprit.

Recording Costs for PC-Fax Card Users — A fax page can be recorded with the computer printer for about one cent per page instead of five cents on a thermal fax unit. Sharpness of recording will suffer, however, unless a high resolution printer is used. With a laser printer, the recording cost can be almost as much as for Group 3 fax recording on thermal paper.

9.3.3 Applications For Fax

Until recent years, fax units were found mainly in big business offices or banks. Fax machines are now everywhere with fax messages treated as written phone calls. A few specific applications follow:

A Philadelphia real estate operator has paid for his new fax unit in the first two weeks. He uses it to present information, including pictures, on properties available for sale to overseas customers.

A Connecticut factory distributor reports his business growth has doubled since he bought a fax unit.

". . . Fax is worth $20,000 a week to me," Vinnie Visciano says in an ad in the *Hartford Courant* for November 27, 1988. The fax cuts the overnight express costs, provides prompt credit check information, and keeps him in touch with his sales staff.

From an NBC news feature on fax:

> For Judge Richard Lang, a fax first. The first court decision on fax. All papers which can be served on an attorney, I have held can now be served by fax.
>
> Paul K., a Detroit caterer, faxed photos of himself to other company fax machines soliciting blind dates. After a Chicago newspaper reported the stunt, he got 1800 replies by fax.

Newsday, the Long Island, New York, daily newspaper, ran a three-page article on fax on July 26, 1988. The following applications for fax were reported:

> Bernie Stebel, a local lawyer, found the only possible way for a honeymoon trip was to take along a fax unit. He said, "It's not that I'm indispensable, but I figured if things happened and I wanted to see the papers, if I had a fax machine there, I could do it." He received or sent 350 pages. Later, he used his fax to complete three transactions in London, Munich, and Amsterdam in just one day. "Had they required mail — or even couriers — it would have taken a week or two."
>
> A Melville, New York, sandwich shop uses fax to receive orders from nearby offices for noontime pickups. Joe Gambo, one of the owners, says, "If I have someone on the phone taking a large order it can take ten minutes while seven or eight people are waiting on line. With fax it comes through, and when a clerk is finished with a customer, he just takes the order off the machine."

The following list of applications gives an idea of the wide variety of fax uses:

- Ad Layouts
- Art Approvals
- Bills of Lading
- Business Letters
- Buy and Sell Orders
- Car Loan Applications
- Circuit Diagrams
- Company News
- Contracts
- Courier Replacement
- Credit Card Authorization
- Credit Forms
- Damage Reports
- Delivery Receipts
- Disk Jockey Record Requests
- Drawings
- Engineering Sketches
- Express Mail Substitute
- Financial Statements
- Football Play Calls
- Installation Orders
- Insurance Applications
- Intercompany Memos
- Invoices
- Junk Mail
- Legal Papers
- Letters
- Libraries, Interlibrary Copies
- Manuscripts and Corrections
- Maps
- Memoranda
- Newsletters
- Newspaper Clippings
- Packing Lists
- Press Releases
- Price Lists
- Product Specifications
- Program Layouts
- Purchase Orders
- Quotations for Bid Backup
- Racing Form Updates

- Repair Information
- Research Paper Copies
- Sales Literature Drafts
- Sales Reports
- Sandwich Shop Orders
- Signed Receipts
- Sketches
- Specifications
- Stock Market Advisories
- Technical Manual Revision Sheets
- Theater Tickets
- Trucking Permits

New models come very rapidly as equipment marketed for a year or so is dropped and features are added or dropped. (See the hints in Section 9.3.2.) Table 9.3.3.1 shows a few typical fax units sold in the United States as of 1989.

The fax units in the table are given only as examples. They are not recommended over others not shown. Brand names include the following: Adler Royal, AEG Olympia, AT&T, Brother, Canon, Cobra Electro, Dynascan, Extel, Fujitsu, Gestetner, Harris/3M, Hitachi, Konica, Medbar, Mita, Mitsubishi, Murata, NEC, Northwest Bell, Nitsuko, Pactel, Panafax, Panasonic, Pitney Bowes, Radio Shack, Ricoh, Sanyo, Savin, Sharp, TIE, Telautograph, Teli, Toshiba, Trans-Lux, and Xerox.

The price spread is rather large as is to be expected. Discounts are usually available on many models. Low-end fax units without paper cutters often sell for less than $1000. Some are in the $500 to $700 range. There is a continuous spectrum of Group 3 fax models from the low-end units with the bare essentials to the high-end units with enhanced features. For convenience, the following paragraphs divide fax units into a few classifications, but, in reality, none exist.

9.3.4 Low-End Fax

This lowest price class Group 3 fax unit is often sold by mail order catalogs, discount chains, and department stores. Unless fax traffic is quite low, the lowest price unit may not be the least expensive one for the business. The low-end unit is compatible with all Group 3 fax units, but some desirable features have been left out to get the lowest cost. The automatic paper cutter for received copies is eliminated due to cost and size considerations. The trade-off is the convenience of a stack of cut sheets *versus* a long strip of recordings falling onto the floor in a pile, until it can be cut up into pages.

The roll of recording paper is 100 feet long or less, instead of more than 300 feet. It not only will run out sooner, but also has more curl from being wound on

Table 9.3.3.1. Group 3 Fax Units in US Market

List Price	*ADV. Price*	*Print*	*Pounds*	*ADF Pages*	*Auto Calls*	*Page Memory*	*Cutter*	*Seconds per Page*
900	628	T	9	0	0		N	
1200		T	12	0	70		N	
1295	686	T	8	5	1		N	
1495	749	T	15	0	9		N	
1495		T	12	5	100		N	
1600	778	T	9	0	30		N	
1695	799	T	9	0	50		N	
1695	880	T		11	32		N	18
1695	889	T	11	1	100		N	
1745	910	T	10	0	100		N	
1749	1029	T	13	5	70		N	
1795		T	8	5	26	3		
1895		T	17	30	100		N	
1995		T	14	15	15	7	Y	
1995		T	14	5	161		N	
1995	1150	T	15	10	0		N	
1995	1068	T	14	10	140		N	
2195	1109	T			70		Y	18
2250		T	16		124	14	Y	12
2695	1499	T	14	15	65	7	N	
3295		T	22	30	136	20	Y	9
4150		T	48	30	132	28	Y	
2995	1889	TT	25	30	82	6	Y	15
3495		TT	31	30	136	20	Y	12
3795		TT		30	100	60	Y	
5995	4399	L		30	114	60	Y	12
8395		L	166	50	150	238		12

ADF — Automatic Document Feeder; T — Thermal; L — Laser (406 x 392 lines/inch); TT — Thermal Transfer.

a smaller core. The modem speed is 4800 b/s instead of 9600 b/s, resulting in higher telephone bills. The less efficient MH coding is used instead of MR, again costing money in telephone charges. If the unit has a document feeder for sending, it handles only a few pages. A time-saving automatic dialer is not offered. Recently, there is a trend toward adding some of these features to the low-end fax units. These then become lower priced almost fully featured fax units.

9.3.5 Desk Top Integrated Fax-Phone

Most Group 3 fax units that do not have a built-in phone will have an RJ-11C jack for a separate modular phone. If one phone line is used for both voice and fax calls, the fax unit may be set to answer after four or five rings. This gives time to answer voice calls before the fax unit seizes the line.

The simplest type of fax unit with a built-in telephone operates much the same as do separate fax and phone units, but takes less desk space, sometimes no more than a larger desk top telephone. Switching between fax and phone is simplified. With ready access to both fax and phone, it is easy to send sketches or memoranda by fax to supplement the phone conversation without leaving the desk. A built-in speakerphone for voice simplifies this procedure. If an automatic calling feature is provided, the same dialer can call either telephone numbers or fax numbers.

A second type of fax-phone unit provides unattended answering of voice and fax calls. One phone number can be used for both purposes. This type is useful for a personal fax unit or for small offices. If a call is received from an automatic calling fax unit, its CNG tone will switch the telephone line to fax use. For a manual call, the calling party can press a touch-tone telephone key to switch into fax. For a voice call, an outgoing message is provided and memory stores the incoming voice message.

9.3.6 Fully Featured (Standard) Fax

The fully featured Group 3 fax unit has many labor-saving and user-friendly conveniences. It has a full-size, 328-foot roll of recording paper, and will handle a high level of traffic. The designs are mature and have been well tested by years of use. Reliability is so high that most users will have no service required in the first two years. Most manufacturers provide similar features, performance, and quality of copy. If an extra feature is added by one manufacturer, almost all others will soon offer the same or a similar feature. The number and type of features provided and advertised as fully featured for a Group 3 fax unit vary, and the price differs accordingly.

The following is a list of features usually found on the standard fax unit. Some are found in low-end fax units:

Fine Resolution — The optional resolution of 203 lines/inch (8/mm) prints smaller type sizes with good clarity.

High Modem Speed — The optional 9600-b/s V.29 modem provides a faster sending speed and thus lower phone line costs.

Automatic Reception — The Group 3 fax unit can be left with power always on and ready to receive 24 hours a day.

Automatic Document Feeder (ADF) — A tray is provided to drop in the documents to be sent. It is not necessary to hand feed one page at a time or to wait at the fax unit while the pages are sent. Thirty-page ADF capacity is common.

Automatic Disconnection — The fax unit automatically disconnects from the phone line after the last page is sent or after a page sent is not received properly. This prevents the fax line from being busy when the unit is not in use.

Automatic Cutter — This very desirable feature cuts the received page to the size of the page sent. Sheet-fed units do not need a cutter.

Journal — A record is made in memory for each document sent or received. This record indicates the date for the journal entry, time of transmission, phone line connection time, identification of the other fax unit, number of pages, identity of other party, and confirmation of receipt. If there were a problem, it would be noted. The journal may automatically print out each day, when there are 32 entries, or on request.

Page Numbering — Each page of the received document is automatically numbered. Some fax units also print "END" on the last page.

Header Print — The date, time, and phone number of sender is printed at the top of each page received.

Automatic Dialing — The phone numbers of fax units called can be stored in memory and used for sending fax transmissions by pressing one or two buttons. Most units allow sending the same documents to groups of fax units by pressing one button for the fax group selected.

Automatic Background Control (ABC) — When sending a page, the same results are obtained for gray or colored backgrounds as for white. The ABC circuit adjusts the background signal level to be the same as white before digitizing it to black-white. This saves the operator trying to guess the proper setting.

Light-Dark-Normal Control — This control allows the operator to compensate for light or heavy markings when necessary.

Front Panel Display — It is common to have an LCD display with 20 or 40 characters to guide the operator in use of the fax unit. The display may also tell what is wrong if the fax unit is not working properly, out of paper, *et cetera.*

Voice Request — The sending operator can set an alarm to ring at the receiving fax unit after transmission is complete. The two parties can then talk without making another telephone call. This feature may be used frequently when a fax unit is first installed. It is seldom needed after users become familiar with fax operation. The receiving station operator can set an alarm for the calling station. This will operate at the end of the page being received. Some fax units will automatically print out a call-back request if the other party does not answer.

Automatic Step-Down Modem — During the initial handshake, a transmission rate of 9600 b/s (if available) is tried first. If the receiving fax unit cannot receive this speed without errors, 7200 b/s is next tried by the transmitter. If this fails, 4800 b/s is tried, and then 2400 b/s, if necessary. This ensures transmission at the highest rate consistent with the quality of the phone line connection.

Compatibility with Group 2 Fax — If the fax called is a Group 2 unit (obsolete design), the page will be sent with Group 2 protocol. This is a feature which becomes less useful each year as the remaining Group 2 units are retired.

Compatibility with Group 1 Fax — There are practically no Group 1 fax units in use, but Group 1 compatibility is also offered on some fax units. This feature may cause more problems than good. After failure to handshake for Group 3 or Group 2, the transmitter will send in six-minute mode, even if not connected to another fax unit. A multipage document "sent" to nowhere can tie up the phone line for a long time.

Copy Mode — A document may be copied by pressing "copy" button. The fax unit then sends to its own printer.

Verification Stamp — Each page sent can be automatically stamped after the fax receiver confirms that it was received.

Photograph Transmission — Pictures can be sent to any Group 3 unit if the fax transmitter encodes the signal to represent gray scale shades by clumps of pels which imitate the photo screening process used by newspapers. Many short coding runs are generated and a single page may take five minutes or so to send.

Polling — This feature commands a called fax unit to send back documents intended for the caller. Fax units that can call fax phone numbers from memory usually can be programmed to call multiple fax units in sequence. A timer allows polling to start at a later time if desired. A fax unit receiving from a large number of other units can automatically call to collect documents left in memory or the document feeder. If polling were not used, each sender would be competing with others to access the fax receiver, leading to many busy signals and frustration. Polling is often done at night when all fax units are unattended. A polling code match between units prevents sending documents to the wrong calling party. Turn-around polling is a feature which allows the calling station to poll the fax unit called automatically after sending documents to it.

9.3.7 Enhanced Features

Page Memory — Most memory fax units will store about 30 pages. The maximum page memory storage available, however, is 1200 pages or more. The user can quickly scan the documents into memory and return the originals to file. Some fax units will sort destinations of the stored documents for the day and batch send all documents for one destination at one time.

CCITT Error Correction Mode — This standardized option automatically retransmits the sections of a page received with errors. The receiving fax unit assembles an error-free page in memory by replacing the errored portions with retransmitted sections. Even if there are no errors, this option takes about 5% longer than standard Group 3 transmission time.

High-Speed Error-Free Mode — This is an NSF option that has more efficient Group 4-like compression and white line skip. This mode operates only with a compatible Group 3 fax unit from the same manufacturer.

Wide Page Sending — A fax unit may have capability of sending wider pages, usually 10 to 12 inches wide. Standard 8½-inch wide pages are also sent without wasting the wider border space. The number of pels per scan line is automatically set as the document feeder guides are adjusted for paper width. When sending wide pages to another fax unit that has 8½-inch wide paper, the original is reduced in size to fit the narrower recording paper width. The transmitter usually throws away pels from the wider page, causing a reduction in resolution.

Automatic Copy Size Reduction — See Wide Page Sending above.

Higher Resolution — An optional resolution of 200 × 400 lines per inch gives better results when sending gray scale images such as photographs. It also increases the sharpness of printed text. The 400 lines per inch is achieved by stepping the page through only 1/400th inch per scan line. The size of the pels may still be 1/200th inch high, giving overlap in the paper feed direction, not true 400 lines per inch. A few fax units have an optional mode of true higher resolution of about 400 lines per inch both horizontally and vertically and with a laser printer. Operation in this mode is only between fax units of the same manufacturer.

Programmed One-Touch Buttons — These allow transmission time, destination, and resolution to be stored and recalled by one button. A different set of transmitting conditions can be associated with different telephone numbers.

Transmission Reservation — When the fax unit is busy sending from memory or receiving, the operator can scan documents into memory along with the transmission details. The original documents may be put back in the files and the pages stored in fax memory will be sent automatically when the unit becomes idle.

Serial Broadcasting — Documents stored in memory can be sent sequentially to many different locations. Delayed sending can be set up to use lower night telephone rates. This is particularly useful for large time zone jumps where no one is at the receiving end during business hours at the sending station.

Different Documents for Different Addresses — Lists can be sorted to have delayed sending of all documents to one destination in one phone call.

Remote (Relay) Broadcast — This system programs receiving fax units to retransmit received fax messages in its memory to another calling group of fax units. One set of instructions can send the same fax information to as many as 6400 fax units.

Telephone Number–Name Storage — Selecting the telephone number by name of the party called simplifies the calling procedure when a large telephone number directory is stored.

Automatic Retransmission — Pages not received successfully can be automatically resent from memory.

Out of Paper Reception — The fax unit continues to receive pages into memory when the recording paper runs out.

Password Operation — Fax units can be set with one or more passwords to ensure that the fax is communicating with authorized parties. This feature is particularly useful in polling. For a station communicating with only a given set of fax units, a password will prevent anyone from sending you junk fax mail.

Confidential Reception — Fax transmissions can be sent directly into memory of the receiving fax unit. A password is needed for retrieval and printout.

Encryption — The facsimile data stream can be encrypted so that it is completely unintelligible to a fax receiver not having the proper code. This prevents the sending of sensitive information to the wrong party by mistake and thwarts unauthorized eavesdropping.

Picture Signal Processing — The received copy quality can be improved by this enhancement. Isolated specks can be removed from the page being sent and aliasing (edge jitter) can be reduced by signal processing.

Proprietary Gray Scale — A nonstandard compression algorithm is used to reduce the transmission time for photographs by reducing the number of short run lengths. This will only work under NSF with another fax unit having the same algorithm.

Interpolation — This technique discussed in Section 6.3 makes standard resolution recorded copy look almost as good as fine resolution.

Automatic Resolution Control — Using NSF, a page will be sent with fine resolution where there are fine details and with normal resolution elsewhere. This saves transmission time compared with sending the whole page in high resolution.

Remote Diagnostics — The supplier's diagnostic service center can call a malfunctioning fax unit and read its diagnostic codes to assist the technician in quickly identifying and solving problems.

Modem Step-Up — This allows the modem to step up to a higher sending rate between pages if the error rate is low and the modem is running at a slow rate. An isolated noise burst during training signal could cause this low speed condition.

User Identity Cards — Credit card sized identity cards can restrict use of a fax unit to card holders. It also lists the user in the journal and can automatically sort out usage costs by department.

Call Status Voice Announcements — Both the sender and recipient are notified of the communication status by voice announcement from the fax unit.

Plain Paper Recording — The higher cost thermal transfer recording and laser recording are used in some fax units for recording on plain paper. The cost of the thermal-transfer fax recorder is less than the laser fax recorder. Cost of thermal-transfer recording materials is about three times that of direct thermal recording. The equipment size of thermal-transfer fax units now on the market is larger than direct thermal recording fax units, but smaller than laser units. Although more plain paper fax designs are being announced, the factors above have limited them to a small portion of the market. If thermal-transfer fax recorders are made in large quantities, both the machine costs and the supplies cost could be much less.

9.3.8 Cellular Mobile Fax

There are about two million cellular phones installed in automobiles and other vehicles (as of 1989). Mobile Group 3 fax units can be mounted in an automobile or other vehicle and communicate with other fax units on the regular telephone network (PSTN) by using one of the 832 channels available in each cell. A special installation in the vehicle is required. Cellular radio uses a four-wire channel instead of a two-wire channel like the PSTN. Most cellular radio telephone units lack provisions for fax and a separate interface box is needed to emulate the PSTN.

The interface box provides ringing current and dc hold current for automatic answering. Dial-tone conversion and tone or pulse dialing would be needed. If the fax unit is not designed to operate on 12-volt vehicular power or its own battery pack, an inverter will be needed for generating ac mains power. The inverter may need automatic switching to prevent its continuous drain on the automobile battery while the fax is not in use.

The range of each cell is only two miles, with automatic switching as the vehicle moves to a new cell. Although switching is not noticeable in voice calls, it may cause problems with fax. With vehicular movement, and even when the vehicle is stationary, switching between cells may occur with temporary signal loss for up to a half second. This will cause a streak about 1/16th inch high across the page. The frequency of cell changing is unpredictable and pages may have no streaks, or up to 10 streaks per page. Switching may also result in a failure message at the fax transmitter, caused by too many received errors. Use of the Group 3 error-correction mode is desirable, but this system has not yet been added to portable fax units.

Encryption of the fax signals is important when sending sensitive information because programmable scanners can be set up to intercept cellular radio transmissions. Signal dropouts could be a serious problem for encrypted signals.

Better voice quality may be achieved in Europe when a new cellular radio system called GSM becomes available. The planned digital data rate of about 12,000 b/s, however, is not suitable for the analog signal delivered for a telephone line by the digital modem of Group 3 fax units. A 9,600-b/s digital interface to GSM for Group 3 fax is being studied. This approach requires that the cellular radio base station provide a real-time D/A conversion at the interface to the PSTN.

The future of mobile fax in the United States is very promising with a new time division multiple access (DTMA) digital system recently endorsed by the Cellular Telecommunications Association. Demonstrations, already made with three subscribers per channel, divided the 30 kHz radio channel into six time slots with a total of 48 kb/s. The technology should allow six to fifteen times as many users without increasing the bandwidth allocated for cellular radio. The new standard is scheduled to be completed by December 1989.

9.3.9 Small Copy A5 and A6 Fax

Some manufacturers expect these smaller A5 and A6 fax units to open up a new home fax market. They certainly are looking for something to pick up the slack when the demand for business fax units begins to taper off with a maturing market. To say that the price will be low enough to generate the high sales figures anticipated is premature. Currently, there is not much interest in Europe or North America.

9.3.10 PC-Fax Cards

The market for PC-fax cards is becoming intense. There are about 40 different cards on the market. Lower prices are attracting more customers. Communicate, Ltd., forecasts placements in Europe of 185,000 in 1988 and 325,000 in 1989.

Surveys show that 70% of the documents sent by fax are generated by a computer or word processor. A PC-fax card instead of a Group 3 fax unit makes sense if you are interested primarily in sending rather than receiving documents, and if they are created on your own PC or network. This can save both time and cost compared with sending by fax.

PC-fax card sellers often point out that much secretarial time would be saved by sending documents directly from the computer rather than printing them for sending on a fax unit. PC-fax sends directly, saving not only the print time, but also the time for transporting the documents to the location of the fax unit, and waiting time when the fax is busy. Unnecessary conversation after completing the work causes additional delays and costs. At the receiving end, someone must call the recipient to pick up the document accruing further costs.

These arguments neglect certain facts. While the computer is sending files it may not be available for anything else. Some PC-fax cards allow receiving at the same time as other work is being done on the computer. If you send fax during much of the day, the computer may not be available for other work.

The document is usually read and approved by the boss of an office before it is sent, and it is not likely to be done from the computer screen at the secretary's desk. A file copy is required as well. This argument is similar to saying that because most of the documents copied are computer or word processor generated, office copiers should be removed and computer printers used instead. A fax machine can be looked at as a remote office copier. PC-fax can be looked at as a remote computer printer. Many customers who buy a PC-fax card already own a Group 3 fax unit. The others often buy one later.

If the hard copy already exists, it can be sent faster and more easily by fax, by anyone, and without the use of a keyboard. PC-fax requires some knowledge and confidence in running a computer. The PC-fax software must be activated and the user must ply his way through screen menus following a number of steps to send the document. Though most secretaries can be trained to use PC-fax, it will not be used by someone who is not familiar with the use of a keyboard.

The cost of the PC-fax board plus a scanner is usually more than a low-end fax unit. With the proliferation of low-cost fax units, it is likely that there is one near the PC and that there will be no waiting.

9.3.11 Junk Fax Mail

Soon the morning's fax mail may be overloaded with unwanted advertising, solicitations, politicians' speeches, and other extraneous material. Unlike the human being who can hang up on unwelcomed calls, the fax unit operates continuously. This uses recording paper, costing five to ten cents a page, and occupies the fax unit so it cannot be used for sending and receiving.

Junk fax mail may have started innocently enough as an extension of normal fax communications for the purchase of paper and supplies. The supplier started sending unsolicited flyers of specials and other fax items. Delicatessens and sandwich shops received noontime pickup orders sent from nearby offices. Then the delis began sending out menus with daily specials, ready for the customer the first thing in the morning. A fax with stored telephone numbers does the calling before normal work hours. The deli owner entered the telephone numbers only once into the fax machine's telephone directory. For sending the daily menus, the pages to be sent are first scanned to enter them into the fax unit's memory. A few key strokes then set up the automatic calling and sending.

PC-fax cards make it much easier to send junk fax mail. The sender does not even need to print out the pages before sending. He is also no longer limited to 100 numbers for sending junk fax mail, but can send to thousands. Separate phone number lists can be kept in the PC for different types of mailings. If the list becomes too large to handle with one PC and one phone line, a fax service like MCI Fax can do it, calling many fax units simultaneously.

As far as undesired junk fax mail is concerned, things have recently taken a disturbing turn. One PC-fax card seller now offers a floppy disk mailing list of other fax users along with his PC-fax card. The package can be programmed to send a mailing directly from the PC to all fax numbers on the list. The floppy disk distribution of specialized fax users means that the sender has all the fax telephone numbers collected for him and only has to transfer the list to his computer hard disk. Even an unlisted number may find its way onto one of these fax floppies for sale. This practice may eventually be self-regulating because users will avoid doing business with those who send junk fax mail.

Some fax users are setting up their Group 3 handshake to drop the telephone connection if it does not contain the proper identification code.

9.4 GROUP 4 EQUIPMENT IN THE MARKETPLACE

The market for Group 4 fax is developing slowly. Table 9.2.2 shows the world Group 4 population at less than 1% of Group 3, with most of the units in Japan. Most of the US Group 4 fax units listed in Table 9.2.2 are from the earlier Federal Express Zap Mail service. At present, Group 4 usage is more that of a computer network communications tool than as a person-to-person unit like Group 3 fax or the telephone.

While Group 3 fax units have instant access to others anywhere in the world via the regular telephone network (PSTN), Group 4 fax units have a difficult and expensive time accessing other Group 4 fax units. This problem is caused by the lack of a universal digital public network. The ISDN promises to provide the communication channel needed for Group 4. Field trials of ISDN have been made in Japan, North America, and most European countries. It is expected to take at least five years before universal access is achieved. In the United States, private digital networks are used with Group 4 fax units adapted to fit the digital network requirements.

Net Express Communications, Inc., offers a public international digital network for connecting to Group 4 fax units in the following countries:

Country	*ROPA*
Australia	OTC
Bahrain	Batelco
Belgium	RTT
Canada	Telecom
France	Nexus
Hong Kong	C&W
Italy	Italcable
Japan	KDD
Kenya	KPT
Korea	DACOM
Phillipines	Eastern Telecoms
Sweden	Televerket
United Kingdom	BTI
West Germany	Bundespost

Connection from the customer office to the network may be at 9.6 kb/s on a dedicated channel or the PSTN. Alternatively, 56 kb/s can be used with a dedicated connection to the network. Communication is at 56 kb/s after reaching the network through 13 access cities in the United States. A protocol converter is usually required. In addition to the above costs, there are monthly minimum usage charges of $500 to $3200 per month. Communication charges per 64,000 bytes are $0.80 within the United States, $1.65 to Europe, and $2.00 to Asia and Pacific countries. This service is designed as a transition to the future ISDN.

AT&T Acunet provides 56-kb/s service, which may be used for Group 4 fax units. Four companies, Canon, Fujitsu, NEC, and Ricoh, are starting to put more effort into marketing Group 4 fax units in the United States in 1989, with some prices as low as $8,000.

9.5 FACSIMILE SERVICES

9.5.1 Bureaufax

Bureaufax is a facsimile service offered by the PTTs in 68 countries for message transmission as a modern replacement for telegrams, mainly using Group 3 fax units. (See Table 9.5.1.1.)

The CCITT has defined the operational procedures in Recommendation F.170. A person wanting to send a *faxgram* uses a standard form for the message. Service is available from public bureau offices in many European countries. Switzerland is an example of a country which has a well developed Bureaufax service.

Table 9.5.1.1. Bureaufax Countries

Antigua	Germany	New Guinea
Argentina	Greece	New Zealand
Australia	Guadeloupe	North Korea
Austria	Guyana	Norway
Bahrain	Hawaii	Panama
Belgium	Hong Kong	Peru
Bermuda	Iceland	Phillippines
Brazil	Indonesia	Portugal
Canada	Israel	Réunion
Cayman	Italy	Singapore
Chile	Ivory Coast	Spain
Costa Rica	Japan	Sri Lanka
Cyprus	Kuwait	St. Pierre and Miquelon
Denmark	Luxembourg	Sweden
Egypt	Macao	Taiwan
Fiji	Maldives	Thailand
Finland	Malta	Trinidad and Tobago
France	Martinique	Turkey
French Guyana	Netherlands	United Kingdom
French Polynesia	New Caledonia	United States

The charges depend upon the length of message and its destination. (See Table 9.5.1.2.)

Bureaufax traffic is rapidly increasing as shown in Table 9.5.1.3. Notice the large percentage increase in traffic per year up through 1985. This corresponds to the customers discovering how useful fax is. The slowdown in growth rate corresponds to a rapid increase in Group 3 fax sales. The service is being modernized with store and forward, and error correction is planned. Bureaufax has been useful in stimulating interest in fax, but is less needed as business owners buy fax and send most of their traffic on regular telephone lines (PSTN).

9.5.2 PTT FAXPOST

FAXPOST is a public fax service primarily for international communication to or from customers who do not have Group 3 fax units. More than 100 countries participate in this service (as of August 1987). Although it can be used within one

Table 9.5.1.2. Swiss Bureaufax Charges
(in Swiss Francs [a])

Destination [b]	*First page*	*Next pages*
National	4	2
Europe	6	3
USA and Canada	8	5
Other Overseas	12	9

[a] Exchange rate (September 1988): SFr 1.00 = $ 0.75.
[b] Delivery by messenger: SFr 4.00 for each message.

country, inexpensive express mail service or Group 3 fax now takes most of this type of traffic. The handling of the fax pages depends on whether the customer at either end has his or her own fax unit. (See Figure 9.5.2.1.)

A received fax copy is delivered to its destination from the FAXPOST receiving station by express mail, usually within four hours. A customer can drop off pages that he or she wants sent at a PTT FAXPOST office where they will be sent to the distant FAXPOST office. Some cities have a pick-up service. Any originals that can be sent by regular Group 3 fax will be accepted for FAXPOST.

The Netherlands, a typical country using FAXPOST, has more than 220 FAXPOST offices. A message within Holland received any time before 6:00 PM will be delivered the same day. The rate is 15 guilders ($7.20) for the first page plus 5 guilders ($2.40) for additional pages.

Charge for service to other countries is from 25 to 45 guilders ($12.00 to $21.60) for the first page, depending upon destination. Additional pages are from 8 to 18 guilders ($3.85 to $8.65). Rates include transmittal sheet and express mail delivery or telephone notification. A new FAXPOST service is "Lucky Courier," which combines a FAXPOST form for your message with a deluxe greeting card.

Sending fax information worldwide can be handled through Bureaufax or FAXPOST. International calls are routed through an Intelpost interface in Miami, Florida. Intelpost is operated by the US Postal Service.

Table 9.5.1.3. Swiss Bureaufax Traffic and Income

Year	*Traffic (in Pages)*	*Increase (in Percent)*	*Income (in Swiss Francs)*
1980	10,700		54,500
1981	17,700	65	84,200
1982	30,900	75	146,600
1983	48,000	55	240,400
1984	99,000	106	378,800
1985	246,400	149	879,500
1986	429,000	74	1,175,100
1987	712,900	66	1,698,500
1988	1,000,000 *	40	
1989	1,250,000 *		
1990	1,500,000 *		
2000	3,000,000 *		

* Estimated.

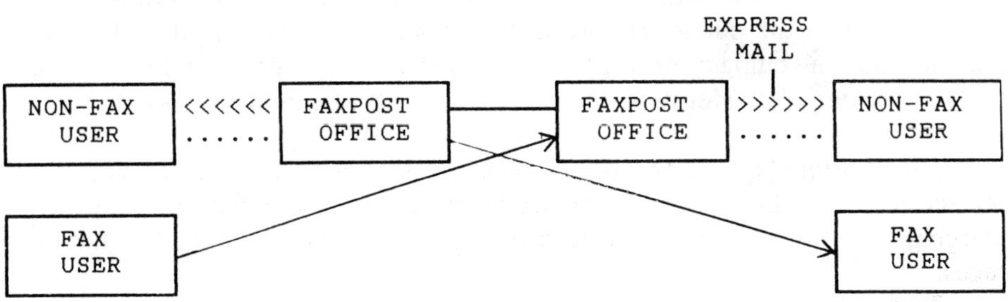

Figure 9.5.2.1 FAXPOST delivery diagram.

9.5.3 Hotelecopy

Hotelecopy is a large fax service bureau operating somewhat like Bureaufax described in Section 9.5.1. Most of the major hotel chains in the United States offer this service. Documents can be sent from one of the participating hotels to any Group 3 fax unit on the regular telephone network (PSTN). Business travelers find this very useful for keeping in touch with the home office. Items needing the traveler's attention can be waiting when he checks in. Copies of reports left behind can be quickly relayed, if suddenly needed. Fax documents can also be exchanged with customers and suppliers for rapid completion of business transactions. Delivery can be made to recipients who do not have fax units by sending to a hotel nearby. The hotel will then deliver the fax by hand. With fax transmission, order numbers and other data will not become garbled.

The charge for sending a document of up to 10 pages is $9 for a hotel guest or $6 if he or she is a member of Hotelecopy's Preferred Member Club. Walk-in service is also available to private individuals and those few businesses who do not have their own fax units, for $12. Wade Helms, EVP and General Manager of Hotelecopy, explains: "One of the great advantages to locating the service in a hotel is that it's open 24 hours a day and it's always safe to walk into."

The service is rapidly catching on. In early 1987 there were 346 hotels in 44 states. By early 1989 the numbers have grown to 1700 hotels in 50 states plus overseas hotels in Puerto Rico, Nassau, Freeport, Aruba, and the Virgin Islands.

9.5.4 Additional Fax Services

Not everyone who wants to use a fax has one yet, but a number of fax services are available to the public. Before these services for instance: At end-of-month deadline time a company with a fax unit might be under pressure to have a confirmation order. A dealer with no fax unit had to find a quick method to send his document.

Many print shops, copy shops, and even stationery stores now offer to send and receive fax. These small businesses invested in Group 3 fax units and started charging about five dollars to send one page and about half that for additional pages.

Some are organized as franchises and will send to other associated shops for pickup by the person needing the received copy. This segment of the fax business is so popular that there is a separate business of selling specialized quarterly telephone directories of fax copy shops and others who will handle the fax copy in another city. There is a service which picks up and delivers fax documents. By having local letter couriers do this work, they avoid the expense of their own delivery system.

Libraries are beginning to offer patrons free use of a fax unit. Calls are charged to the user's telephone number or credit card.

Other firms are selling franchises for self-service public fax booths. With easy access at any time, the customer does not have to wait for a shop to open in the morning or worry about finishing his document before the shop closes. Some fax booth systems allow the use of a credit card.

Resellers of long-distance telephone line service are trying to compete with low-cost transmission services that bypass long-distance phone companies. One firm, as an example, offers a fax switching service that sends the customer's fax document from its store-and-forward computer to one or many Group 3 fax units worldwide. For example, the rate to England is $1.29 for the first minute and $0.90 for each additional minute. Another entrepreneur is selling shared use of his 800 number to public facsimile service operators.

9.5.5 DHL Worldwide Express

This is an overnight electronic document delivery service using 400-line/inch fax units to send customers' documents. When operating at 9600 b/s on the PSTN it sends CCITT Test Document No. 1 in 66 seconds. The plain paper laser fax unit made by Canon is compatible with Group 3 and Group 4 fax units.

The system is designed to use X.25 protocol on the Net Express Communications, Inc., international packet switching network and will send the same document in 10 seconds at a 64-kb/s rate. The company has operating agreements with the national telecommunication administrations of 13 foreign countries. Access to the US network is through 13 city nodes. The Net Express network can also handle Group 4 transmissions.

9.5.6 MCI Fax

MCI Fax is a premium connection service for MCI customers to send by Group 3 fax to US and overseas Group 3 users connected to the PSTN. A dedicated, digital, optical fiber network designed for 9600 b/s is used for domestic and international facsimile transmission. Connection to the office or home is provided by the local exchange carrier, which uses its own copper telephone pairs and charges for their use. Office-hour rates are about 25 to 35 cents per minute in the United States and higher on overseas calls. Fax rates are higher than for a regular MCI voice call. Other services such as transmission to many fax users and store-and-forward are offered. The customer buys his own Group 3 fax unit from other sources.

MCI Mail-to-Fax Dispatch is a separate electronic mail service (discussed in Section 9.5.7 below).

9.5.7 Electronic Mail with Group 3 Fax Delivery

Many public electronic mail systems (EMS) now use Group 3 fax units for message delivery. Messages in low-order ASCII characters are received from subscriber data terminals, PCs, mainframes, or telex. The system then converts the message to Group 3 fax format and sends it to the customer's Group 3 fax unit over the PSTN. A message can be distributed to hundreds of recipients with a single call. The EMS systems can send, but cannot receive, fax messages.

Western Union EasyLink is an EMS service that costs $25 a month. Documents generated on a PC may be sent to Group 3 fax units at a rate of 55 cents for the first half page and 35 cents for additional half pages.

Syscom has a worldwide fax delivery service called WORD*FAX. Any communicating terminal, such as a PC, may be used to send messages for delivery via Group 3 fax units. The rate is $2.00 per page plus telephone costs.

MCI Mail Fax Dispatch (MCI owns RCA Globecom and Western Union International) costs 60 cents for the first half page and 40 cents for additional half pages in the United States.

Among other companies providing this service in the US are:

- AT&T Mail
- Electronic Courier Systems
- Gemfax New Facsimile Switching Network

Some of the EMS systems can also receive Group 3 fax messages for relay to multiple fax addresses, as MCI Fax can (Section 9.5.6 above).

9.5.8 The Future

Now that fax has finally been recognized as essential to business communication, we ask: What next? Will it keep pace with newer technology, or fall behind and be replaced? Looking toward the future, consider the following:

Group 3 Stand-Alone Fax — Group 3 fax with a configuration similar to the current one will be in great demand for many years. More features can be expected, including higher speeds and higher resolution.

Group 4 Fax — The future for extensive large use of this standard is unclear. Digital fax for networking, store-and-forward, and image databases is in the offing, and Group 4 may fill this need. Development of a viable mixed mode with teletex is unlikely, but computer mixed mode is probable. Group 4 will have very little impact on Group 3.

Color Fax — The technology exists today, but costs are much higher for color. The business market has a limited need for a color fax unit, similar to needs for a color office copier.

Transmission Speed — T-3 networking with business customer access at 45 Mbits (equivalent to 672 voice channels) is starting. According to Stephen B. Weinstein, Manager of Multimedia Communications Division, Bell Communications Research, Morristown, New Jersey,

> With or without video distribution, it became clear in 1988, that optical fiber to the home is now an economic reality.*

Optical broadband ISDN draft standards have now been recommended by the CCITT and completed standards are expected in 1992. Fujitsu has just demonstrated a 1.8-Gb/s (equivalent of 24,000 voice telephone circuits) optical fiber subscriber loop conforming to the CCITT draft standard.

What does this mean to the future of fax? Tom Valovic, Senior Editor of *Telecommunications,* states,

> Group III and IV facsimile will be major contributors to the huge appetites for bandwidth that many companies are already developing.†

No doubt the future is bright, and the opportunity exists for the better fax mentioned above and for other types of fax not yet conceived.

* S. B. Weinstein, *IEEE Spectrum,* Vol. 26, No. 1, January 1989, p. 43.

†T. Valovic, Editorial, *Telecommunications,* Vol. 26, No. 1, January 1988, p. 6.

Glossary of Facsimile Terms

AM-PM-VSB — Vestigial sideband amplitude modulation-phase modulation such as used for CCITT Group 2 facsimile.
Analog Facsimile — That form of facsimile which uses *analog facsimile signals* between the *facsimile transmitter* and the *facsimile receiver.* Areas of the *original* with a density between black and white may be recorded as gray shadings.
Analog Facsimile Signal — A facsimile signal which represents a continuum of scanned *original* image densities by a corresponding range of signal values.
Aperture Distortion — The distortions in resolution, density, or shape of the recorded image caused by the shape and finite size of the scanning and recording apertures.
Available Line — In *analog facsimile,* that portion of the scanning line which can be used specifically for *picture signals* or *recording.*

Bandwidth Compression — A technique to reduce the bandwidth needed to transmit a given amount of facsimile information in a given time or to reduce the time needed to transmit a given amount of facsimile information in a given bandwidth.
Baud — The number of changes in signal state per second in a digital signal sent by a modem. A baud may contain four or more bits as in a CCITT V.29 modem.
Bit — The contraction for *binary digit,* the smallest amount of information in a binary system, a 0 or 1 condition.
Black Signal — In *analog facsimile,* the signal at any point in a *facsimile system* produced by the *scanning* of a maximum *density* area of the *original.*

Calibration Scan — A preliminary scan, usually at low resolution, to determine the density range of the *original.*
CCITT — Consultative Committee on International Telegraph and Telephone. An intergovernmental advisory organization of the International Telecommunication Union, which recommends worldwide communication standards including those for *facsimile.* See Groups 1, 2, 3, and 4.

Color Facsimile System — A *facsimile* system that produces the *recorded copy* in more than one color. Typically, a full color image is produced by sending images for three primary colors.
Compatibility — Matching *facsimile transmitter* and *facsimile receiver* characteristics that permit the receiving of acceptable facsimile copy.
Compression Ratio — In *digital facsimile,* the ratio of the total bits used to represent the *original* to the total number of encoded bits.
Computer Generated Facsimile Signal — A signal generated by programmed processing of information for image recording on a local or remote *facsimile receiver.*
Continuous Tone Image (Analog Gray Scale Image) — An image in which each resolvable element may be represented by one of a continuous range of tones.
Contouring — Density step lines in *recorded copy* resulting from quantization of an original image that has observable gray shadings between adjacent quantization intervals.

Definition — See *Resolution.*
Dielectric Recording — See *Electrostatic Recording.*
Digital Facsimile — That form of *facsimile* in which densities of the *original* are sampled and quantized as a digital signal for processing, transmission, or storage.
Direct Recording — That type of *recording facsimile* in which a visible *recorded copy* is produced without subsequent processing.
Document — A set of one or more pages which can be transmitted as a unit.
Drum Speed — The revolutions per minute of the transmitter or recorder drum. See *Scanning Line Frequency.*

Electrographic Recording — See *Xerographic Recording.*
Electrolytic Recording — Recording with signal-controlled current through an electrolyte in the recording paper, depositing metallic ions to produce a mark.
Electronic Shading — An electronic method of compensating for variations in sensitivity of individual sensors of a sensor array or variation in illumination of copy being scanned. This may be done by correcting the analog signal from each sensor sample under control of stored digital information.
Electrophotographic Recording — See *Xerographic Recording.*
Electrosensitive Recording — Recording with an electrical signal which passes directly into the *record medium.*
Electrostatic Recording — *Recording* by means of a signal controlled electrostatic field. (Note: A toner is required to make the image visible.)
Electrothermal Recording — See *Thermal Recording.*
End-of-Line (EOL) — In *Group 3 digital facsimile* systems, a sequence of digital symbols introduced at the end of a scanning line to establish synchronization of decoding and for error detection.

Facsimile — The process by which a document is scanned, converted into the

electrical signals, transmitted, and recorded or displayed as a copy of the *original.*
Facsimile Copy — A *recorded copy* of an *original* produced by a *facsimile recorder.*
Facsimile Receiver — The apparatus employed to translate *picture signals* from the communications channel into a *facsimile copy* of the *original.*
Facsimile Recorder — That part of the *facsimile receiver* which performs the final conversion of electrical *picture signals* to an image of the *original* on the *record medium.*
Facsimile Signal — See *Picture Signal.*
Facsimile Test Chart — A document used for evaluating the performance of a *facsimile* system.
Facsimile Transceiver — An apparatus that sends and receives *facsimile signals. Full-duplex* models can send an *original* at the same time that a different *recording* is being made. *Half-duplex* models can only send or receive at a given time.
Facsimile Transmitter — The apparatus employed to translate the *original* into *picture signals* suitable for delivery to the communication system.
Fax — An abbreviation for *facsimile.*
Fingerprint Facsimile — *Facsimile* equipment used to transmit fingerprint cards. (Note: Existing systems send 8 × 8-inch cards at 192 lines per inch.)
Flood Projection — The optical method of *scanning* in which the *original* is floodlighted and the *scanning spot* is determined by a masked portion of the illuminated area.
Framing — Adjustment of the received picture to a desired position in the direction of the recording paper progression.

Ghost — In *analog facsimile,* a spurious image resulting from echo, envelope delay distortion, or multipath reception.
Gray Scale Characteristic — The relation between the density of the *recorded copy* and the density of the *original.* (Note: This term sometimes means a relation between the facsimile signal amplitude and the density of the *original* or *recorded copy.*)
Group 1 — *Analog facsimile* equipment per CCITT Recommendation T.2. It sends an A4 or 8½ × 11-inch page in 6 minutes over a voice grade telephone line using frequency modulation with 1300 Hz corresponding to white and 2100 Hz to black of the *original.* (Note: Because North American 6-minute equipment uses 1500 Hz white and 2400 Hz black, it is not compatible with *Group 1 equipment.*)
Group 2 — *Analog facsimile* equipment per CCITT Recommendation T.3. It sends an A4 or 8½ × 11-inch page in 3 minutes over a voice grade telephone line using 2100-Hz AM-PM-VSB.

Group 3 — *Digital facsimile* equipment per CCITT Recommendation T.4. It sends an A4 or 8½ × 11-inch page typically in one-half minute over a voice grade telephone line.
Group 4 — *Digital facsimile* equipment per CCITT Recommendations T.5 and T.6, which uses Public Data Networks and their procedures for essentially error-free reception. It may also be used on the Public Switched Telephone Network with an appropriate modulation process.
Grouping — Periodic error in the spacing of *recorded lines.*

Halftone Image — An image that has been converted from a continuous tone image into a two-tone image while retaining the appearance of a continuous tone image.
Handshaking — An exchange of signals (called "control procedures") between the *facsimile transmitter* and *facsimile receiver* to verify that facsimile transmission can proceed, to determine which specifications will be used, and to verify reception of the documents sent.
Horizontal Resolution — The number of *picture elements* per inch (or millimeter) in the direction of *scanning* or *recording.*

Index of Cooperation — In *analog facsimile,* the product of the drum diameter times the number of lines per unit length or the product of the *total line length* times the *lines per inch* (millimeter) divided by π.
Ink Jet Recording — That type of *recording* in which ink particles are directly deposited upon the *record sheet.*

Jitter (in Facsimile) — Irregular error in the position of the *recorded spot* along the recorded line. (Note: This is noticeable on the recording of a vertical line.)

K Factor (in Modified Read Coding for Group 3 Facsimile) — The number of facsimile scanning lines in a set used for coding. At the most, $K - 1$ lines are coded two-dimensionally to limit the disturbed area in the event of transmission errors. In CCITT Group 3, $K = 2$ for 3.85 lines/mm and $K = 4$ for 7.7 lines/mm. In Group 4, $K = \infty$.

Landscape Mode — The mode of scanning lines across the longer dimension of an *original.* See *Portrait Mode.*
Line Feed Error — See *Grouping.*
Lines per Inch (or Millimeter) — The number of scanning or recording lines per unit length measured perpendicular to the direction of scanning.
Line-to-Line Correlation — The correlation of image information from scanning line to scanning line. Useful for two-dimensional coding, e.g., *modified read.*

Magnetic Recording — *Recording* by means of signal-controlled magnetic field.
Maximum Keying Frequency — The frequency equal to one-half the number of *picture elements* per second.
Mobile Facsimile — *Facsimile* equipment used in vehicles.
Modified Huffman (MH) Coding — A one-dimensional run length digital scheme of coding white and black runs, where the shortest length code words represent the most probable run lengths. MH is used by Group 3 facsimile.
Modified Read (MR) Coding — A two-dimensional optional digital coding scheme for Group 3 facsimile. (Note: MR provides an improved transmission speed over *modified Huffman* coding.)
Modified Modified Read (MMR) Coding — A two-dimensional coding scheme for Group 4 facsimile. (Note: MMR provides improved transmission speed over *modified read coding.*)
Moiré — The spurious pattern in the reproduced copy resulting from interference beats between two or more sets of periodic structures in the image or system.
Multipath Distortion — In *analog facsimile,* distortion of the recorded copy caused by a radio propagation phenomenon, which results in signals reaching the receiving antenna by two or more different path lengths. The corresponding spread in signal arrival time causes *jitter, ghost,* or smear of the received image for *analog facsimile* or errors for *digital facsimile.*

Newsphoto Facsimile — *Facsimile* equipment used to transmit photographs for newspaper or magazine publishing.

Original — A page which is transmitted by *facsimile.*

Pagination System — *Facsimile* equipment is used as part of a system which produces page layouts, including text, screened photographs, and graphics, to produce masters for publishing newspapers or magazines.
Pel — A *picture element* that contains only black-white information (no gray shading). See *Pixel.*
Phasing — In *analog facsimile,* adjustment of the *recorded copy* to the proper position on the *record sheet* in the direction of the *recording spot.* (Note: An "out-of-phase" condition results in a split image on the *record sheet.*)
Photographic Recording — Recording by the exposure of a photosensitive surface to a signal-controlled light beam or spot. See *Electrophotographic Recording.*
Picture Element — The smallest area of the *original* sampled and represented by an electrical signal. See *Pel* and *Pixel.*
Picture Signal — A signal resulting from the *scanning* process or an electronically generated equivalent.

Pixel — A *picture element* that has more than two levels of gray scale information. See *Pel.*
Pixel Interpolation — Generation of additional unscanned *pixels* by logical comparison of nearby scanned pixels to simulate increased resolution. Interpolation may be one- or two-dimensional.
Portrait Mode — The mode of scanning lines across the shorter dimension of an original. (Note: Groups 1, 2, 3, and 4 use *portrait mode.*) See also *Landscape Mode.*
Press Facsimile — *Facsimile* equipment that transmits full newspaper or magazine pages to produce masters for publishing at remote printing plants.

Quantizing Levels — In a *digital facsimile* system, the number of different gray steps representing a *continuous tone* image. See *Contouring.*

Raster Scanning — The method of *scanning* in which the *scanning spot* moves along a network of parallel lines, either from side to side or top to bottom.
Record Medium — The physical medium on which the *facsimile recorder* forms an image of the *original.*
Record Sheet — The medium used to produce the *recorded copy.* The *record medium* and the *record sheet* may be identical.
Recorded Copy — A hard copy of the *original* produced by facsimile.
Recording — The process of converting the *picture signal* in a *facsimile receiver* to an image on the *record medium.* See *Direct Recording, Electrolytic Recording, Electrosensitive Recording, Electrostatic Recording, Ink Jet Recording, Magnetic Recording, Photographic Recording, Thermal Recording, and Xerographic Recording.*
Recording Spot — The image area corresponding to a *picture element* formed at the *record medium* by the *facsimile recorder.*
Redundancy Reduction — Coding for elimination of redundant information in the *picture signal* to reduce the amount of information needed for transmission or storage. (Note: The amount of *redundancy reduction* will vary with the information content of the *original.*) See *Compression Ratio.*
Resolution — A measure of capability to delineate picture detail.

Scanner — That part of the *facsimile transmitter* which systematically translates the *densities* of the *original* into a signal waveform.
Scanning — The process of successively analyzing the *densities* of the *original* according to a predetermined pattern.
Scanning Direction — Normal direction is from left to right and top to bottom of the *original,* as when reading a page of print.
Scanning Line Frequency — The number of scanning lines per unit of time. (Note: This is equivalent to *drum speed* in some mechanical systems.)

Scanning Line Length — See *Total Line Length.*
Scanning Spot — The area on the *original* viewed instantaneously by the photosensor of the *scanner.*
Signal Contrast — In *analog facsimile,* the voltage ratio expressed in decibels between *white signal* and *black signal.*
Skew — (1.) The deviation of the *recorded copy* from rectangularity due to asynchronism between *scanner* and *recorder.* (2.) Angular misalignment of the *original* from the paper feed direction. (3.) Deviation of the angle of the *scanning line* or *recording line* from perpendicular to the paper path.
Soft Copy — That form of *facsimile* which displays the received image on a cathode ray tube or similar screen.
Spot Projection — The optical method of *scanning* in which the *scanning spot* size is determined by the illuminated area of a spot which moves across the *original.*
Spot Speed — The speed of the *scanning* or *recording* spot within the *available line.* (Note: This is generally measured on the *original* or on the *record sheet.*)
Subject Copy — See *Original.*
Synchronizing — The maintenance of the proper position of the *recording spot* while it is writing to produce an undistorted *recorded copy* of the *original.*

Thermal Recording — That type of *recording* which is produced principally by signal-controlled thermal action. (Notes: Direct *thermal recording* involves direct imaging on the *record medium. Thermal transfer* recording involves heat from a thermal print head transferring marking from a carbon ribbon or overlay sheet to another sheet, forming the *recorded copy.* Some recorders produce full color recordings by successively overwriting different colors in the same area.)
Thermosensitive Recording — See *Thermal Recording.*
Throughput Time — The time to transmit and receive one page. The timing begins with the first operator, electrical, or mechanical event associated with transmission, and ends when the *recorded copy* is available. For a multiple-page document, the time is measured for continuous transmission, dividing the total time by the number of pages. (Note: Time to establish the communication circuit is not included in this definition.)
Tone — A value of optical density for a single color.
Total Line Length (TLL) — The TLL is equal to the *spot speed* divided by the *scanning line frequency.* In *digital facsimile,* the TLL is equal to the *pels* per scan line divided by the number per millimeter.
Transmission Time — The time for sending a single page (i.e., elapsed time between the start of *picture signals* and the detection of *end-of-message signal* by the facsimile receiver).

Usable Line Length (ULL) — *Total line length* minus the "dead sector" (such as the clamp bar width). See *Available Line.*

Weather Facsimile — Various classes of *facsimile* equipment used to transmit weather charts, cloud cover images, or weather radar images.

White Signal — In *analog facsimile,* the signal at any point in a *facsimile system* produced by the *scanning* of a minimum *density* area of the *original,* or the electronically generated equivalent.

Xerographic Recording — Recording by action of a light spot on an electrically charged photo-conductive insulating surface where the latent image is subsequently developed with a toner.

Bibliography

Albert, Arthur Lemuel, *Electrical Communication,* Second Edition, New York: John Wiley and Sons, 1940, pp. 3–6.

Berger, Meyer, *The Story of the New York Times,* New York: Simon and Shuster, 1951, pp. 408–413.

Costigan, Daniel M., *Electronic Delivery of Documents and Graphics,* New York: Van Nostrand Reinhold, 1978, pp. 1–21.

Costigan, Daniel M., *Fax: The Principles and Practice of Facsimile Communication,* Philadelphia: Chilton, 1971, pp. 1–11.

Jones, Charles R., *Facsimile,* New York: Murray Hill, 1949, pp. 1–23.

Reynolds, F. W., "A New Telephotograph System," *Electrical Engineering,* September 1936, pp. 996–1007.

Ranger, Richard H., "Transmission and Reception of Photoradiograms," Institute of Radio Engineers, New York, June 3, 1925.

Service Bulletins, "History of Facsimile," Times Facsimile Corporation, Vol. 4, No. 1–Vol. 6, No. 10.

"Facsimile Edition Hailed as Journalistic Milestone," *Times Talk,* Vol. 10, No. 1, September 1956, pp. 1–4.

Weinstein, Stephen B., "Broadband Communications," *IEEE Spectrum,* Vol. 26, No. 1, January 1989, p. 43.

Index

Acoustic coupling, 18, 180
Algorithm, 23–24, 68, 78, 115–133, 137, 146–148, 150, 152, 157–158, 160, 195
Aliasing, 195
Amorphous silicon, 103–104
AP (Associated Press) 10–15, 17, 180
Architecture, 28–29, 87, 89, 131, 160
ASCII, 71, 80, 82–84, 94, 119, 121, 127, 131, 181, 206
AT&T, 9–12, 16, 127, 181, 189, 200, 206
Automatic background control (ABC), 103, 192
Automatic document feeder, 46, 185–186, 192

Bain, Alexander, xi, 2–3, 5
Bakewell, Frederick, 3
Bandwidth, 39–40, 159, 207
Baud, 24, 29, 38
Bell, Alexander Graham, xi
Bit-mapped, 80, 95
Broadcast, 11, 15, 195
Bureaufax, 200–204

Canon, 178, 189, 200
Carlson, Chester F., 110
Carrier frequency, 10, 40
Carterphone, 18
Casselli, Giovanni, 3–4
CCD, 28–29, 101–104
CCITT, 15, 18, 21–24, 25–26, 31, 38, 40, 49, 62, 71, 75, 78–80, 83, 165–176, 184, 194, 200, 205, 207
CED, 42–44

Cellular phone, 83, 196
CFR, 42–44
CIS, 103–106
CNG, 79, 191
Coding, 2, 23–24, 26, 29, 31–38, 184, 191, 193
Compatibility, 17–18, 21, 25, 27, 49, 78, 83, 85, 94, 97, 193, 197
Color fax, 154–160, 206
Conditioning, line, 10, 12, 25
Contouring, 137
Cooley, Austin G., 11–12, 14
Copier, 27, 103, 109, 111, 163, 178–180, 186, 198, 206
Courier, 27, 181–183, 187–188, 202, 205–206
CSI, 46–49
Cutter, 184, 189, 192

Dacom, 23–24
d'Arlingcourt, Ludovic, 4
DCN, 42–44
DCS, 42–44
Dealers, 180, 184
Delay distortion (*see* Envelope Delay and Distortion)
Density, 163, 165
Deskfax, 11, 180
Dialer, automatic, 27, 46, 186, 191
DIS, 42–44, 46–49, 75
Distortion, 9–10, 18, 28, 40, 102, 123, 135, 137, 140, 150–153, 163–166
Dither, 62–69, 140
Drum, 1, 3–6, 10, 14, 111–113, 158, 180

Echo, 10, 49
Electrolytic recording, 2–3, 5
Electronic Industries Association (EIA), 17–18, 21, 163
Electrophotography, 110
E-Mail, 83
Encryption, 82, 195, 197
Envelope delay, 9–10, 40
EOL, 32, 42–44, 52, 57
EOP, 42–44
Equalization, 10, 28, 38
Error checking, 32, 37, 42, 49, 52–57, 62, 92–93
Error correction, 42, 46, 49, 52, 57, 62, 79, 83, 93, 194, 197, 201
Error-free, 25, 43, 62, 79, 84, 194
Europe, 14, 24, 87, 159, 177, 180–181, 197, 199–200

Faxgram, 200
FCC, 29, 197
Fork, tuning, 4–5, 10, 14
Frequency response, 40
Frequency standard, 4–5, 13

Gemfax, 206
Gifford, Walter, 12
Graphic Sciences, 18, 21
Gray scale, 11, 63, 65–67, 135–154, 163–164, 193–195
Group 1, 18, 21, 26–27, 84, 85, 193
Group 2, 21, 23, 25–27, 84, 85, 193
Group 3, 2, 13, 25–84, 85–86, 94, 96–98, 101, 103, 107, 109, 111, 115–116, 123, 131, 157, 166–176, 180–181, 184–185, 187, 189–194, 196–202, 204–207
Group 4, 37, 65, 69, 78–79, 83, 85–99, 106, 111, 113, 131, 137, 159, 181, 194, 199–200, 205–207

Halbrooks, Chuck, 14
Handshake, 27–28, 38, 44, 49
Helms, Wade, 204
Hotelecopy, 204

IBM, 169, 181
IEEE, 163–165, 181, 207
Intelpost, 202
Interoperability, 18
Interpolation, 119, 123, 127, 195
ISDN, 79, 85–86, 92, 96–98
ISO (International Standards Organization), 86, 89, 91, 99, 131, 137
Ives, Dr. Herbert, 10

Japan; Japanese, 17, 23, 94, 133, 152, 160, 177–181, 199–200
Jitter, 31, 40, 195
Journal, 192, 196
Junk fax, 186, 195, 198–199

Korn, Dr. Arthur, 5–7

Laser fax, 11, 27, 80, 82–83, 94, 110–111, 187, 194, 196, 205
LED (light-emitting diode), 105–106, 111–113
Loss, telephone line, 9–10

Magnavox, 18, 180
Matsushita Graphic Communication Systems, 104, 178
MCF, 42–43, 45–48
MCI Fax, 199, 205–206
Messagefax, 15
Modem, 17–18, 24, 25–31, 38–43, 49, 62, 69–71, 79–84, 85–86, 93, 97, 184, 191–193, 196–197
Modified Huffman (MH) code, 28–32, 38, 52, 57, 62, 71, 80, 184, 191
Modified read (MMR) code, 68, 96, 119, 127
Modified read (MR) code, 28–30, 38, 42, 52, 57, 62, 71, 119, 121, 127, 131, 166, 184, 191
Morse, Samuel, xi, 1–2, 32
Morse code, 2

National Communication Systems (NCS), 169
NEC (Nippon Electric Company), 178, 189, 200
Net Express, 200
The New York Times, 11–12, 14, 16–17
Newsphoto, 6, 10–11, 31
Noise, telephone line, 10, 18, 40, 37, 104, 121, 137, 141, 150, 196
NOMDA (National Office Machine Dealers Association), 180

Nonstandard,
NSC, 46
NSF, 45–49, 184, 194–195
NSS, 45–48
operation, 77–78

Ochs, Adolph S., 11
OCR (optical character recognition), 86, 127
Optical fiber, 93, 205, 207
Optics, 13, 103–106, 112–113
OSI (Open Systems Interconnect), 85–94, 97

Packet switching, 84, 205
Password, 82, 195
Pattern recognition, 86, 127
PC-Fax, 80–84, 86, 101, 111, 186–187, 197–199
Phasejitter, 40
Photofacsimile, 11
Photosensitive, 102, 111–113
Photosensor, 1, 7, 31, 101–106, 158, 166
PIN, 42, 45
PIP, 42, 45
Policefax, 15
Polling, 46, 186, 193–195
Portable fax, 11–16, 83, 197
Pressfax, 15, 17
Printing, 17, 29, 63, 75, 82, 86, 94, 107–113, 121, 157–159, 166, 184, 198
Protocols, 38, 40, 46, 52, 71, 78–79, 83–84, 86–94, 193, 200, 205
PSTN (Public Switched Telephone Network), 17–18, 24, 25, 27–29, 38, 69, 71, 79, 83–84, 85–86, 92, 96–98, 119, 185, 196–197, 199–201, 204–206
PTT, 29, 180, 200–202

Quantizing, 31, 136–137, 140–153

Radio, 5, 15, 93, 163, 196–197
Rapifax, 23–24
Readability, 25–26, 29, 127, 166
RCA (Radio Corporation of America), 9, 206
Receiver, fax, 3, 7, 14, 17, 28–31, 37–38, 40, 42–43, 46, 49, 52, 57, 62, 65, 75, 77–78, 82, 99, 107, 119, 121, 123, 137, 141, 150, 154, 159, 193, 195
Recommendations (CCITT), 21–23, 25, 31, 38, 40, 46, 49, 62, 75, 78–79, 83, 85, 89–99, 159–160, 165–166, 200
Recorder (*see* Receiver)
Redundancy reduction, 23, 29, 31, 52, 115, 123, 135, 137, 140, 143, 148, 152
Reliability, 63, 92, 109–110, 112, 179, 183, 191
Resolution, 6, 10–11, 17, 21, 25–26, 29–31, 37, 71, 75, 79, 82–83, 94–95, 101, 106, 107, 123, 127, 131, 150, 154, 158–160, 163–166, 169, 184, 186–187, 192, 194–196, 206
Retrieval, 80, 86, 131, 160, 195
Rod lens, 104, 106, 112
Rotating mirror, 111–112
RTC, 43, 47
RTN, 42, 45
RTP, 42, 45
Run length, 23, 29, 31–32, 116, 166, 195
RS-232 interface, 38, 69–71, 84, 185

Satellite, 11, 17, 158
Scanner, 3, 8, 28–29, 31, 71, 82, 85–86, 101–106, 158, 166, 186, 197–198
Scrambling (*see* Encryption)
Security, 80, 82, 89, 185
Selfoc, 104
Sensor, 1, 7, 31, 101, 103–106, 158, 166
Seven-layer model (*see* OSI)
Shannon, Claude E., 31
Sharpness, 65, 82, 104, 187, 194
Shonnard, John, 14
Sidebands, 40
Skew, 1, 3, 31
Soft copy, 80, 131, 160
Standards, 13–15, 17–18, 21–23, 25, 83, 85–99, 131, 159, 163–176, 180, 207
Stenafax, 15
Stix, Gary, 181
Store-and-forward, 84, 85, 205–206
Storage, 27, 31, 80, 82, 84, 86, 94, 109, 119, 147, 154, 160, 194–195
Stylus, 1, 3–4, 11
Sulzberger, Arthur H., 11–12, 14, 16
Switzerland, 200
Synchronization, 1, 49

Telautograph, 181, 189
Telecopier, 180
Telegram, 11, 24, 80, 200

Telegraph, 1–3, 5–6, 8, 15, 24, 32, 180
Telephoto, 10, 14
Teletex, 24, 78–79, 91–92, 94–96, 98, 206
Telex, 24, 27, 78–79, 83, 91, 181, 206
Thermal recording, 27, 107–109, 186–187, 196
Thermal-transfer recording, 109–111, 196
Times Facsimile Corporation (TFC), 13–16
TSI, 46–49
TSR, 80
TSS, 46

US Army Air Force, 14–15, 23
US Army, 15, 23, 180
US Navy, 12, 15

Valovic, Tom, 207

Wall Street Journal, 17
Weatherfax, 15
Weber, Donald, 23
Weinstein, Stephen B., 207
Western Union, 9, 11, 24, 180, 206
Wirephoto, 10, 17

Xerography, 110–111
Xerox, 18, 21, 180–181, 189